FIRESIDE SERIES

Volume 3, No. 3

PARALLEL LIFETIMES:

FLUCTUATIONS
IN THE
QUANTUM FIELD

ISBN # 1-57873-115-1

JZK Publishing
A Division of JZK, Inc.

P.O. Box 1210
Yelm, Washington 98597
360.458.5201
800.347.0439
www.ramtha.com
www.jzkpublishing.com

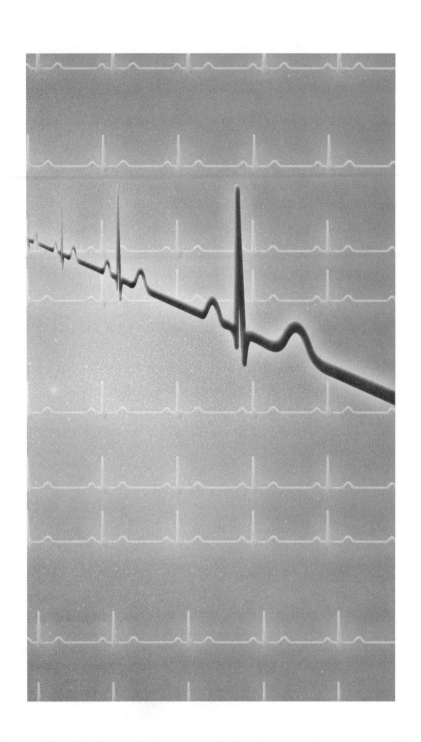

These series of teachings are designed for all the students of the Great Work who love the teachings of the Ram.

It is suggested that you create an ideal learning environment for study and contemplation.

Light your fireplace and get cozy. Have your wine and fine tobacco. Prepare yourself. Open your mind to learn and be genius.

FOREWORD TO THE NEW EDITION

The Fireside Series Collection Library is an ongoing library of the hottest topics of interest taught by Ramtha. These series of teachings are designed for all the students of the Great Work who love the teachings of the Ram. This collection library is also intended as a continuing learning tool for the students of Ramtha's School of Enlightenment and for everyone interested and familiar with Ramtha's teachings.

In the last three decades Ramtha has continuously and methodically deepened and expanded his exposition of the nature of reality and its practical application through various disciplines. It is assumed by the publisher that the reader has attended a Beginning Retreat or workshop through Ramtha's School of Enlightenment or is at least familiar with Ramtha's instruction to his beginning class of students. This required information for beginning students is found in *Ramtha, A Beginner's Guide to Creating Reality*, revised and expanded ed. (Yelm: JZK Publishing, a division of JZK, Inc., 2000), and in *Ramtha, Creating Personal Reality*, video ed. (Yelm: JZK Publishing, a division of JZK, Inc., 1998).

We have included in the Fireside Series a glossary of some of the basic concepts used by Ramtha so the reader can become familiarized with these teachings. We have also included a brief introduction of Ramtha by JZ Knight that describes how all this began. Enjoy your learning and contemplation.

CONTENTS

FIGURES

Introduction to Ramtha
by JZ Knight

"In other words, his whole point of focus is to come here and to teach you to be extraordinary."

You don't have to stand for me. My name is JZ Knight and I am the rightful owner of this body, and welcome to Ramtha's school, and sit down. Thank you.

So we will start out by saying that Ramtha and I are two different people, beings. We have a common reality point and that is usually my body. I am a lot different than he is. Though we sort of look the same, we really don't look the same.

What do I say? Let's see. All of my life, ever since I was a little person, I have heard voices in my head and I have seen wonderful things that to me in my life were normal. And I was fortunate enough to have a family or a mother who was a very psychic human being, who sort of never condemned what it was that I was seeing. And I had wonderful experiences all my life, but the most important experience was that I had this deep and profound love for God, and there was a part of me that understood what that was. Later in my life I went to church and I tried to understand God from the viewpoint of religious doctrine and had a lot of difficulty with that because it was sort of in conflict with what I felt and what I knew.

Ramtha has been a part of my life ever since I was born, but I didn't know who he was and I didn't know what he was, only that there was a wonderful force that walked with me, and when I was in trouble — and had a lot of pain in my life growing up — that I always had extraordinary experiences with this being who would talk to me. And I could hear him as clearly as I can hear you if we were to have a conversation. And he helped me to understand a lot of things in my life that were sort of beyond the normal scope of what someone would give someone as advice.

It wasn't until 1977 that he appeared to me in my kitchen on a Sunday afternoon as I was making pyramids with my husband at that time, because we were into dehydrating food and we were into hiking and backpacking and all that stuff. And so I put one of these ridiculous things on my head, and at the other end of my kitchen this wonderful apparition appeared that was seven feet tall and glittery and beautiful and stark. You just don't expect at 2:30 in the afternoon that this is going to appear in your kitchen. No one is ever prepared for that. And so Ramtha at that time really made his appearance known to me.

The first thing I said to him — and I don't know where this comes from — was that "You are so beautiful. Who are you?"

And he has a smile like the sun. He is extraordinarily handsome. And he said, "My name is Ramtha the Enlightened One, and I have come to help you over the ditch." Being the simple person that I am, my immediate reaction was to look at the floor because I thought maybe something had happened to the floor, or the bomb was being dropped; I didn't know.

And it was that day forward that he became a constant in my life. And during the year of 1977 a lot of interesting things happened, to say the least. My two younger children at that time got to meet Ramtha and got to experience some incredible phenomena, as well as my husband.

Later that year, after teaching me and having some difficulty telling me what he was and me understanding, one day he said to me, "I am going to send you a runner that will bring you a set of books, and you read them because then you will know what I am." And those books were called the Life and Teaching of the Masters of the Far East (DeVorss & Co. Publishers, 1964). And so I read them and I began to understand that Ramtha was one of those beings, in a way. And that sort of took me out of the are-you-the-devil-or-are-you-God sort of category that was plaguing me at the time.

And after I got to understand him, he spent long, long

moments walking into my living room, all seven feet of this beautiful being making himself comfortable on my couch, sitting down and talking to me and teaching me. And what I didn't realize at that particular time was he already knew all the things I was going to ask and he already knew how to answer them. But I didn't know that he knew that.

So he patiently since 1977 has dealt with me in a manner by allowing me to question not his authenticity but things about myself as God, teaching me, catching me when I would get caught up in dogma or get caught up in limitation, catching me just in time and teaching me and walking me through that. And I always said, "You know, you are so patient. You know, I think it is wonderful that you are so patient." And he would just smile and say that he is 35,000 years old, what else can you do in that period of time? So it wasn't really until about ten years ago that I realized that he already knew what I was going to ask and that is why he was so patient. But as the grand teacher that he is, he allowed me the opportunity to address these issues in myself and then gave me the grace to speak to me in a way that was not presumptuous but in a way, as a true teacher would, that would allow me to come to realizations on my own.

Channeling Ramtha since late 1979 has been an experience, because how do you dress your body for — Ram is seven feet tall and he wears two robes that I have always seen him in. Even though they are the same robe, they are really beautiful so you never get tired of seeing them. The inner robe is snow white and goes all the way down to where I presume his feet are, and then he has an overrobe that is beautiful purple. But you should understand that I have really looked at the material on these robes and it is not really material. It is sort of like light. And though the light has a transparency to them, there is an understanding that what he is wearing has a reality to it.

Ramtha's face is cinnamon-colored skin, and that is the best way I can describe it. It is not really brown and it is not really white and it is not really red; it is sort of a blending of that. And he has very deep black eyes that can look into you and you know you are being looked into. He has eyebrows that look like wings of a bird that come high on his brow. He has a very square jaw and a beautiful mouth, and when he smiles you know that you are in heaven. He has long, long hands, long fingers that he uses very eloquently to demonstrate his thought.

Well, imagine then how after he taught me to get out of my body by actually pulling me out and throwing me in the tunnel, and hitting the wall of light, bouncing back, and realizing my kids were home from school and I just got through doing breakfast dishes, that getting used to missing time on this plane was really difficult, and I didn't understand what I was doing and where I was going. So we had a lot of practice sessions.

You can imagine if he walked up to you and yanked you right out of your body and threw you up to the ceiling and said now what does that view look like, and then throwing you in a tunnel — and perhaps the best way to describe it is it is a black hole into the next level — and being flung through this tunnel and hitting this white wall and having amnesia. And you have to understand, I mean, he did this to me at ten o'clock in the morning and when I came back off of the white wall it was 4:30. So I had a real problem in trying to adjust with the time that was missing here. So we had a long time in teaching me how to do that, and it was fun and frolic and absolutely terrifying at moments.

But what he was getting me ready to do was to teach me something that I had already agreed to prior to this incarnation, and that my destiny in this life was not just to marry and to have children and to do well in life but to overcome the adversity to let what was previously planned happen, and that happening including an extraordinary consciousness, which he is.

Trying to dress my body for Ramtha was a joke. I didn't

know what to do. The first time we had a channeling session I wore heels and a skirt and, you know, I thought I was going to church. So you can imagine, if you have got a little time to study him, how he would appear dressed up in a business suit with heels on, which he has never walked in in his life.

But I guess the point that I want to tell you is that it is really difficult to talk to people — and perhaps someday I will get to do that with you, and understanding that you have gotten to meet Ramtha and know his mind and know his love and know his power — and how to understand that I am not him, and though I am working diligently on it, that we are two separate beings and that when you talk to me in this body, you are talking to me and not him. And sometimes over the past decade or so, that has been a great challenge to me in the public media because people don't understand how it is possible that a human being can be endowed with a divine mind and yet be separate from it.

So I wanted you to know that although you see Ramtha out here in my body, it is my body, but he doesn't look anything like this. But his appearance in the body doesn't lessen the magnitude of who and what he is. And you should also know that when we do talk, when you start asking me about things that he said, I may not have a clue about what you are talking about because when I leave my body in a few minutes, I am gone to a whole other time and another place that I don't have cognizant memory of. And however long he spends with you today, to me that will maybe be about five minutes or three minutes, and when I come back to my body, this whole time of this whole day has passed and I wasn't a part of it. And I didn't hear what he said to you and I don't know what he did out here. When I come back, my body is exhausted and it is hard to get up the stairs sometimes to change to make myself more presentable for what the day is bringing me, or what is left of the day.

You should also understand as beginning students, one thing that became really obvious over the years, that he has shown me a lot of wonderful things that I suppose people who have never gotten to see them couldn't even dream of in their wildest dreams. And I have seen the twenty-third universe and I have met extraordinary beings and I have seen life come and go. I have watched generations be born and live and pass in a matter of moments. I have been exposed to historical events to help me to understand better what it was I needed to know. I have been allowed to walk beside my body in other lifetimes and watch how I was and who I was, and I have been allowed to see the other side of death. So these are cherished and privileged opportunities that somewhere in my life I earned the right to have them in my life. To speak of them to other people is, in a way, disenchanting because it is difficult to convey to people who have never been to those places what it is. And I try my best as a storyteller to tell them and still fall short of it.

But I know that the reason that he works with his students the way that he does is because also Ramtha never wants to overshadow any of you. In other words, his whole point of focus is to come here and to teach you to be extraordinary; he already is. And it is not about him producing phenomena. If he told you he was going to send you runners, you are going to get them big time. It is not about him doing tricks in front of you; that is not what he is. Those are tools of an avatar that is still a guru that needs to be worshiped, and that is not the case with him.

So what will happen is he will teach you and cultivate you and allow you to create the phenomenon, and you will be able to do that. And then one day when you are able to manifest on cue and you are able to leave your body and you are able to love, when it is to the human interest impossible to do that, one day he will walk right out here in your life because you are ready to share what he is. And what he is is simply what you are going to become. And until then he is diligent, patient, all-knowing, and all-understanding of everything that we need to know in order to learn to be that.

And the one thing I can say to you is that if you are interested in what you have heard in his presentation, and you are starting to love him even though you can't see him, that is a good sign because it means that what was important in you was your soul urging you to unfold in this lifetime. And it may be against your neuronet. Your personality can argue with you and debate with you, but you are going to learn that that sort of logic is really transparent when the soul urges you onto an experience.

And I can just say that if this is what you want to do, you are going to have to exercise patience and focus and you are going to have to do the work. And the work in the beginning is very hard. But if you have the tenacity to stay with it, then one day I can tell you that this teacher is going to turn you inside out. And one day you will be able to do all the remarkable things that in myth and legend that the masters that you have heard of have the capacity to do. You will be able to do them because that is the journey. And ultimately that ability is singularly the reality of a God awakening in human form.

Now that is my journey and it has been my journey all of my life. And if it wasn't important and if it wasn't what it was, I certainly wouldn't be living in oblivion most of the year for the sake of having a few people come and have a New Age experience. This is far greater than a New Age experience. And I should also say that it is far more important than the ability to meditate or the ability to do yoga. It is about changing consciousness all through our lives on every point and to be able to unhinge and unlimit our minds so that we can be all we can be.

You should also know that what I have learned is we can only demonstrate what we are capable of demonstrating. And if you would say, well, what is blocking me from doing that, the only block that we have is our lack to surrender, our ability to surrender, our ability to allow, and our ability to support ourself even in the face of our own neurological or neuronet doubt. If you can support yourself through doubt, then you will make the breakthrough because that is the only block that stands in your way. And one day you are going to do all these things and get to see all the things that I have seen and been allowed to see.

So I just wanted to come out here and show you that I exist and that I love what I do and that I hope that you are learning from this teacher and, more importantly, I hope you continue with it.

— JZ Knight

WHO IS THE GREAT ARCHITECT OF CREATION?

Greetings, my beloved people. I am indeed most pleased to be with you at this wondrous and grand event called Assay.[1] And Assay means test.

To what I am, such an event offers, for that which I am, a greater unbridled ability to teach more expansive concepts. It allows those concepts to move outside of that which is termed your frame of thinking and expand, open the doors to your mind, allows me to do more, to teach you more.

This platform, indeed, always will I rush to be a part of because in this school, in all the years that it has been fundamentally in place, has been a school based on one pure concept: that you are God. But to say that and to simply say, "Now I am enlightened, I have the answer" would assume that you would understand and know utterly and totally the nature of God. The school started with this premise and your, as human beings, relationship to that profound statement.

This school has had many years of developing — seven years in fact — of developing disciplines, disciplines that were taught, engaged, retaught, reengaged. This was the seven-year period of learning the mechanism of consciousness and energy and how it would prove itself in this school, that this treasured statement of Behold God, the treasured understanding of self in terms of consciousness and energy and mind, would not be lost upon philosophical grounds that are empty but a pristine, a sterling opportunity that the disciplines allowed an application to see a remarkable part of yourself. Maybe it was only a day, maybe it was a half a day, maybe it was seven days but, nonetheless, a watermark on beginning

---

1 Assay I, June 2-10, 2001, held at Ramtha's School of Enlightenment in Yelm, Washington.

the concept of change and daring to move outside of the box based on that one remarkable thing that you did that gave you insight to your possibilities.[2]

What is time in the light of all eternity? How much can you spend on investigating the unknown within your own brain? How much time can you spend on investigating and understanding the brain's connection to the body, the body's connection to the brain? How much time should you spend on understanding the vortex whirling around you of people, places, things, times, and events? How much time should you spend on understanding how they got there? And in this vortex we would assimilate this in terms of our own whirlwind, indeed our own life. How much time would you need or do you think you should spend on understanding that?

In the light of all eternity, it is when we err that we rush to judgment in philosophical terms and carry a postulate in our mind that is so complete in its intellectual or psychological or religious connotation that if you can quote that, there is this sense in you that you now have an understanding of the spiritual life. No one should ever rush to any certitude of judgment that suggests that because you can repeat the words should mean the end of it, that in this little phraseology is what I believe in. This little box, this little phrase in my mind, this is my spiritual life. This is what I believe in and that is my philosophy. It is my religion. It is my psychological understanding of the meaning of my life, is this little box. And we bring it forward when the time is right, when we want to be spiritual, and we push the buttons of that neuronet. And here come those little sentences. They come up and you can espouse them and say them, but then when they are gone they melt away, and now you are back within the whirlwind of your emotion. You have done your duty to reveal your mind.

It took a long, long time in terms of eternity for God — who so passionately is in love with humanity — it took so long for that great passion of God experiencing itself to finally

2 See *Ramtha's School: A Look Within,* video ed. (Yelm: JZK Publishing, a division of JZK, Inc., 2000).

be as far-flung as space itself. But God's passion, God's bliss, superconsciousness, its passion for humanity — each human being, because in each human being is its mirror and its potential the greatest — it took in terms of eternity quite an extraordinary journey of such a passionate being to develop out of its own creation towering concepts in invisible worlds when space, as we see it and indeed know it today, was only a potential; that the only thing upon the vista and the platform of reality was each engagement of creator, its creation of its own consciousness and energy, an unfolding out of each of those consciousness-and-energy ideas that ideas took form. They took form. It is when a barren landscape, vast, into the nothing, started to take on a pictorial dimension, those ideas were being formed, experiences occurring.

And the great architect built the most wondrous and fantastic planes, did so at no urgency — did so at no urgency — for what should be known of urgency? And this great being, this supraconscious, all-in-all, powerful creator to barren mind, barren consciousness, developed from it mind, dimensionalized, a barren vastness into singular beauty. And it kept on and on and on and on. And each level was only because that this being had reached a parameter of collective, self-realized thoughts that — such astounding beauty in the dimensions — that this God, consciousness and energy, had created out of its own self, ran into the concept that could not be maintained on that level because the concept was a quantum leap into a new idea, and it leapt into the mind of God.

And there was then yet another barren plane, another barrenness on the flow of the river of consciousness and energy. And there this magnificent being went about with exalted realization of bringing into form that great wisdom and profound concepts that had so built within the self, called mind, into now a much different phase, a much different time, a much different structure, that these structures would be heavier, that these structures could be broader, that time itself could begin to expand. And so the highway of a dream to its eventual experience was getting longer and more beautiful.

This God that I tell you about is your God. It took an unimaginable amount of time, but time — time — was not the law but used in the law for dimensionalizing thought. This God, after developing all those levels, had developed and took eons to develop the perfect carriage. Once the Earth was germinated and seeded and the seeds were coming right out of the consciousness that is on this plane, that is distinctive to this plane, that when all was made out of the consciousness of this time, distance, space, and mass, and the Gods creating ideas, bringing them into mass — the great gardener of Eden can't pick a flower unless he becomes as the flower — the great gardener of the Earth, after bringing all of this into manifestation, could not create a simple single-cell organism that could be called life until that God had become it analogically. And that would be called the breath of life, instilling in that organism not only its idea from its God but leaving that idea in there as a mind, an analogical mind, that could then develop that single-cell organism's DNA and its eventual social consciousness.

These Gods can create a flower. Every flower, every insect, every bacteria, every virus — everything — is alive because it is given life through the breath of the analogical experience, the God who analogically becomes it and instills within this lifeless idea a living breath, and that living breath is called mind. It has its own mind, and that mind has unlimited potential. So if the Gods have created all of this, then what did they do when it came time to pluck the flower out of the garden with no hand to do so, no nose to smell, no eyes to perceive the true color of its DNA? And God created man and woman. You all have heard all of the stories, and for over twenty years I have elaborated on these stories in great and marvelous detail.[3]

3 See *Human Civilization, Origins and Evolution*. Part I of *A Master's Reflection on the History of Humanity* (Yelm: JZK Publishing, a division of JZK, Inc., 2001).

Suffice it to say, however, what we bring to this Assay is the history of that creation of our human self. We bring to this Assay you, in bodies that were formed more than ten and a half million years ago, DNA. You bring to today the concept of the great Holy Spirit called God that produced itself a viable vehicle, and yet the viable vehicle in the earlier stages had the sharp architect creator always adjusting, creating it to a degree that it could walk — the great gardener — could walk in its garden and grasp a flower, a body that was created out of the same substance that flowers are created out of, in a consciousness that the flower comes from, the Earth comes from, the water of the Earth comes from, for this body was created of this consciousness and it vibrates at the same rate of speed. It is no different than bacteria; it vibrates at the same rate of speed. Why? That the God who inhabits this body cannot only continue to be an analogical creator, an endless analogical creator, but with the brain that has the capacity for that God to unfold from its very origins into the quintessential body of its now great frontier, into a body that not only can it enjoy the life it set in motion, but it has a body now that can — once it has created the dream — it has a body now that it can use to touch the dream, to smell the dream, to hear the dream, to taste the dream, to smell the dream.

What brings you here to this Assay is because you have forgotten what your mission is. You have forgotten the tools you have, the mechanism you have, and you have forgotten that you are part and parcel responsible for the levels of creative fertility. As the ages and the procession, the procession of stars that pass, the ages and their procession pass, people on the Earth and most other planetary systems were susceptible to catastrophic changes in the equilibrium of space.

The first great Gods that were here were so mindful. They understood how to take care of that. But every successive generation has lost that knowledge and the catastrophes have destroyed them, have destroyed their brains, their bodies, their flesh. And then the new seeds that come were further and further behind the eightball, more lost and more degeneration.

As you sit here today before me, you were once unimaginably beautiful. You were once unimaginably imaginably. You were once the great caretakers of a self-loving God that even the flower and even the insect of which that God created brought the nature of bliss to a God who had created something so lovely in its greatest accomplishment, a body in which that loveliness could be appreciated.

You stand here, sit here, on the precipice of some dramatic changes, not only geologically speaking but consciously speaking. Indeed you are here today to begin to relearn what you know but in a much more in-depth field. If you are to save yourselves, then you have to remember with what faculty you have the ability to do that with. And unless you know those faculties, you are the blind leading the blind and into the ditch do you go.

The beautiful nature of God has been lost, because as the human experience becomes thicker in its emotional expression, it loses its creativity. And only through magnetism and gravity does it draw into its whirlwind, into its life, that which supplements its feeling character, its emotional character. The great God of your being is not that God is dead; it is you who are dead. And in this whirlwind, you have perpetuated this through generation after generation to where you are so heavy in mass — so heavy, so caring about that mass, but only to the ends — not that the mass could create, but to leaping into a maturity of the mass so you could participate in the river of adult emotions without ever having to create anything except keeping your body fit enough to have the experience.

As we sit here today and we look out at you, you would be hard-pressed — hard-pressed — to remember when your life took a turn, and that the great creative force came back into your mind, came back into your inspiration to where you have left behind the whirlwind, and that the cause of our life here is not to get hung up in our emotions or even to perpetuate the idea but to have come back to that extraordinary throne in our brain to where the great architect, you, long forgotten, lives in seclusion, secrecy. The great architect, the great God, has done little because it has been asked to do little in your life. It has given you life in your brain, in your cells, in your DNA, a DNA that was clay made off the banks of this river called consciousness and energy.

Understanding what I have just said brings us back to the God in the garden, finally touching that flower and maybe even picking it. What extraordinary sensation, that God plant only that flower? Would that be the only flower? Would the only flowers that would ever rule the meadows of high places be only that flower?

You haven't understood that on this precipice that you are on, you can either continue in the whirlwind with your boxed notes of spiritual retort, philosophical explanation, or you can come back again and learn the secret teachings, ancient teachings, learn them not as if they were outside of the frame of your capacity but that you are the frame of its capacity.

All these years in school have been a remarkable and beautiful journey, a journey that is indeed compelling, a gravity of passion, and a hatred of bitterness. But the school sustains through the deepest, darkest hells of a person's mind to the jail that they are locked up in and held captive by their emotions. This school, in the broadest sense of the term, has endeavored in its tiny presence in eternity to address you since you were that great God on a campaign of self-realized love, the creative thrust of making known the unknown, building on a barren landscape greater concepts of mind and continuing on in its passionate assault of knowing itself.

Now to understand that which is termed the nature of that sublime creator, who could so passionately be in love with the wondrous body/thinking mechanism that was able to rationalize, formulate memory, paradigms — beautiful, you know, like the computer can assume the proportions of that which is lost and give you a visual picture of what could be recovered in memory — God has created the most exquisite body, and that is you, and your ability to think, to have conscious meaning in that thinking, to learn how to rationalize, dimensionalize, and propagate thought.

This tiny school, this tiny law, has been here to address every aspect of yourself that has thickened and become heavy, and that what should always have always been known to you is now a quest for the remarkable or maybe the natural. But if it is no longer natural, it may be remarkable indeed how you paint the picture.

So who is the giver of life? God. Is God alive? No, God gives life. God is eternal. That slid off your back earlier, but it is a profound statement and a clue to understanding you.

Now subtly over the years in saying that you are God, you can only imagine how many encroachments of the divine that I have spoiled, soiled, and outraged — that at the vault of my heaven, and that I consider that the perfect temple of God is the human body, because I know it is — and that I also understand that in nature, the greatest temple to God is the vault of heaven, backdrop of forever, upon glistening stars that only hint to us of extraordinary life that live within them, live within the rays in which they shine so brightly, and that God is replicated in the Earth in the form of nature. And whoever is an Observer of nature will understand the profound secrets that no holy book could ever write. But simply the greatest message ever written is observation and participation in understanding the rite of the great platform in which the great God, so loved in human form, has made nothing into something, and that every something houses the breath of life of that analogical climax that gave it life from that God, and that God is everywhere, and that every stone, every pebble, every granule of sand, every leaf, every color, every drop of hydrogen, every cloud, every rain, every ray of sunlight is within itself an exclusive platform in which many Gods in many forms brought into being by being exclusively those things and giving to it life that decorated the stage of our imminent drama.

THE AKASHIC RECORD, THE MIND OF GOD, AND THE QUANTUM FIELD

That most beautiful and sublime and totally loving being, that I arrested your attention early with in explaining in the most impoverished of ways, lives in you. But you are only life. And your attitudes that we talked of yesterday, those attitudes are partly responsible in God's greatest secretary, Madam Soul. The soul is the recorder of unfinished business, the tallier in which in the mind of God each subject's achievements are added to this fluid mind that the ancients used to call the Akashic Record, but all it means is space. And we know it today in a much more sophisticated term called the quantum field, and its spiritual name is the mind of God.

The Akashic Records, one of those occult terms, but you know all that it means? Space. So let's get rid of the flowery occult stuff. An Akashic Record meant space with knowing — space with knowing — an ether, a subtle fluid that contained every thought, every action, every deed, fulfilled and unfulfilled, by every generation, not only here but every flora and fauna and every bacterial generation. Anything that God gave life to, its life is captured in this space. Moreover, it is just not people here on the Earth. What about the people in the Earth? What about the people above the Earth? What about the people in the Milky Way? But in space, what is called the mind of God in no-time, you who have lived always are imprinted upon this structure in all thoughts, all words, all deeds, all intentions — everything, a living record — along with everything else. Not only does that space involve you but it also involves eleven and a half billion other planets in the Milky Way that support life, just like this one, in degrees of involution and evolution.

If God then only needed one of those[4] in which to shape new paradigms, then God was a very mobile entity whose dreams of what-if were more powerful than what-was. Moreover, every analogical thought then that ever existed, in essence, is the mind of God. Every flower that waxes and wanes in spring and summer was given life through such a climax, and then God moved away from the flower so the flower now with its breath of life could be its own being and indeed intrinsically sponsor and populate itself. If God is always analogically present, the flower never forms. But it had to be becoming and then moving away by the grace of God to allow what God has created now to be a living thing, a living and breathing thing.

Look around us. All this has been brought from nothing into something by an analogical deity who didn't linger here, who didn't linger in the woods but made the lilies of the field, who didn't linger in the lilies of the field but made the insects golden and lovely, who kept moving, because to stay analogically in something is to never complete it and to give it life. And the God of power has moved out, and there then becomes a living and breathing thing. And who could say then that the life force of plants and rocks, bacteria and animals is not divine? It is all divine, because within every living thing there was once a God who dreamt it, became it, and moved out of it, and left with it that divine principle of eternity.

Now what egregiousness you show when you hold onto people, places, things, times, and events because they serve you climactically. How egregious of you to cling to people only under the notion of what they can do for you. You have never been the relationship and then moved out of it to allow the relationship to be its own living thing, like the gardener. You have never cultivated something with such enduring patience, care, attention — in essence, focus — that the triumphantness of its stupendous celebration would bring within itself climax, fruition, and an overabundance of joy,

4 Analogical thought.

that who is to say that that which springs from the heart and the soul is not more powerful than that which springs from the loins.

You hold onto things that you experienced, guided by the soul, but to this day have never given them their solitary life and moved on. And without that, and with that, you have not heard the voice of God any longer in your head. You have only heard what you need to do, what you need to do, what you need to do, what you got to do, what you need to do, what you have got to do, because you are unwise people because you have never finished what you keep saying you need to do.

So God, primary consciousness, you might say went about its great mission to make known the unknown and indeed interacting with all that was it, that it loved, but went on and walked away and kept creating in yonder valley. The God is mining creation there — and yet these Gods, that is their mission — but what they have left behind is never behind, because the life of each creation, the life being the analogical expression that God gave as consciousness and awareness to that form that allowed it to be animate, we call that simply life. But in life and its awareness, whether seeds of its own regeneration, the doingness of created life brings back to the creator the great gift. And the great gift, we can call that in terms of simplistic saying, is that all of these lifeforms, these ideas from the mind of God really didn't come from the mind of God. They came from God as consciousness and energy, and it was God who gave them life. But their life, the continuation of their life, the propagation of their ability to idealize concepts, even mass to mass, to create, to experience, that from life's activity generated a form of thought, that we would know clearly today that it is not the brain that creates mind, nor is it consciousness and energy — although it is the substance to which mind would spring — that mind is a phenomenon that is the result of consciousness and energy on the brain, its ability to move within its river of consciousness upon its vista and to perpetuate and to create within its capacity, that that living form's mind, mind, will flow back into the God mining new life in the valley, and that is called the gift back to God. It is called essentially the mind of God. God has no mind. Mind is the reporting card of the adventures of life, and then God's mind is determined by what has been created and then its free will to live.

David Bohm's Implicate Order and Quantum Unpredictability

Now David Bohm, as he is called, an entity that lived in the last century, understood that there were particles that did not react to the stabilization of the atom, that the atom was a stable force here, or a particle force, or an atomic force in which light was made off of, packets of light — collision of positrons and electrons created an explosion; one of each made one photon — and that in this realm that the only action that could actually be understood was the decay off of certain atoms from some elements. Are you following me? All you have to do is look at, buy, an elemental chart: gold, iron, lead.

However, he saw and knew within this realm, this plane, this level of consciousness, that there were particles that did not stay within the realm. He would say that these particles blinked off and on: one would appear here, go out, and appear over there. Are you with me? Now he called this realm, going up with your big arrow,[5] the implicate order. He didn't know about this.[6] He just called it the implicate order. David Bohm did not know about these levels of consciousness and energy. Everyone concludes in science there is a zero place, but they don't

5 The fourth, fifth, sixth and seventh levels of consciousness and energy. See Fig. B: *Seven Levels of Consciousness and Energy* in the glossary.
6 The various levels of consciousness and energy beyond visible light taught by Ramtha.

understand the nature of that zero space relative to the space and particles they are studying. Are you with me? So imagine for a moment that this is the only level that Bohm was studying, because this is also the level of light. So particles, they would flash in and flash off. So he said those particles come from an implicate order and they unfold into the explicate order.

So what did David Bohm not know about these particles that would blip in and blip out, just flashed in, flashed out? And some that he saw as a particle — "Oh, my God, there it is; it is that purple one." Purple one flashes in, turns off and flashes over here. Was that the same particle? Was it? Are each of these purple particles the same particle? (Audience: No.) Can I tell you why? So if this purple represents, just for understanding, a slab of wood, and there were multiple slabs of wood, can you make a house out of one, an altar to God? Can you make a box? Can you make a bridge? Can you make a road? Can you make an elephant? Can you make a bird? What can't you make out of this slab of purple? So now in David Bohm's implicate order, he was seeing the same particle flash, disappear, and reappear somewhere else, and he concluded it was the same particle. But maybe it was a slab of wood and then a house or an altar to God. How many of you understand?

Now why then was not this background, this static field, consistent? Why were there virtual particles flashing in saying, "Hello, I am here in the explicate; now I am gone." Why were they doing that? Why weren't they static like they have been in all other atmospheres?

Now how many of you understand why, according to Bohm's — now listen carefully — implicate order, that to the explicate that the languoring of those particles was not consistent? How many of you understand now? Do you? Because those particles belonged to the implicate order of how many planes? (Audience: Four.) The implicate "mortis" — well, that is true for some of you; it is really a dead thing — but the implicate order then is actually comprised of four other distinct levels of consciousness and energy and particle reality. And what gives them the context of being an implicate order is because they cannot be sustained in the realm of light that demands that sustenance and sustaining be particles with spin, and they don't have any of that because they are not polarized. I mean, you ought to be so happy to understand such a great mystery, the definition of the implicate order. I mean, if you know that, we are going to understand how to manifest real well. Do you understand?

So what are these segregated particles that physicists study that only blink on once and then they blink on somewhere else, but they think it is the same particle? And so what they are doing is they are measuring in the quantum rough background, the rough field here, they are endeavoring to measure if that particle has a spin, and if they understand it does, then they can calculate its mass and they can calculate its mass from the velocity of where it will appear again. But guess what? What further complicates the problem is to focus on one particle. And then they see this one, and the moment they see this one they have lost this one, but they remember it mathematically. So they can plot the weight and the velocity of that particle in a quantum field.

Now imagine this is a human big fat brain like yours but a little bit more educated. And so this educated brain started this process in linear physics — and not only Newtonian physics; physics means to study particles — so linear physics was to study the nature of the decay of any one atomic structure in terms of time, distance and space. Nuclear physicists study the action of the nucleus of atomic structures, and subatomic physicists, quantum physicists, are studying particles outside of the atomic structure. Got that? So now astrophysicists would be studying large bodies in space and their potential movement. In other words, astrophysicists would be studying the Milky Way and clusters of stars within it and their momentum and their gravity field and how they affect the other stars around them. Now what is similar to astrophysicists and quantum physicists is that they are both studying particles. One just happens to be larger than the other one.

So now Bohm then understood something, along with his guru. His guru said this is the Akashic Record. All of you to some degree have been influenced by ignorant gurus and ignorant knowledge, and the knowledge is to be able to read the Akashic Records. Well, Akashic in ancient Hindu, coming from the description of Sanskrit, only means space. But when David Bohm looked at the implicate/explicate order, he saw space, and his guru said, "Yes, but that that you peer upon is the ether and which we call the Akashic Record." David Bohm parted with his guru in this context. David Bohm said, "If this is the Akashic Record, then I must spend the rest of my life explaining the karma of those virtual particles that flash in and flash out and which I cannot measure their distance, their mass, their property, their velocity." And what do you think the guru said? "Trust me."

David Bohm survived his guru but he survived without his reputation intact. And the only thing that survives from his truly brilliant mind and colored superstition was that he was endeavoring to understand this realm in terms of karma, in terms of the Akashic Record, and which David Bohm himself could only understand with his theoretical, mathematical mind, concepts of particles that were here to affect this reality. And indeed he concluded that there are particles that are never manifested in this form, and those particles he would call later virtual particles because they come in for a moment and then disappear, are never constant.

His dying wish — so we now know the Akashic Record didn't save him — his dying wish was to understand what particle does the Observer focus upon to bring that elusive virtual particle into a constant in the frame of light and matter, and where was the Observer in bringing this into constitutional viability. And David Bohm, at the end of his days, had to part from his guru because to the level of knowledge the guru had, had no understanding about becoming the tiny. The guru only saw it as the Akashic Record of information. "Yes," he said, "that may be true. But what is the nature of any particle that constitutionalizes the unfolded field? Tell me then, as in Morse code. Read me the dots and dashes of these particles if I am to conclude they are the Akashic Record of all life." And in the end, David Bohm lost his prestigious reputation as a physicist because he was misled by a controlling guru, that the guru himself, so blatantly uneducated not as to understand particles as infinite life and David Bohm not to understand that particles were life, and so he leaves us with — he leaves us with — the implicate order and the explicate order on the bridge we call the light,[7] which in those terms will constitute the body, the very large, very massive body that we hold today.

Gross matter is made of atomic structures, atoms, bound not necessarily with the same atoms. In other words, the atoms that make up a piece of wood are a whole library of atoms that have various degrees of chemistry in them. For example, no one can look at a piece of wood and say, ah, this is a piece of fiber, because if we analyze the fiber down to its molecular basis, we will find that there is no such thing as atomic wood, that atomic wood is made up of many atoms that give us the illusion of wood. But the chemistry that makes up wood — the sap, the coagulation, the carbon levels, the H_2O levels — if we were to remove any of those from a piece of wood, it would no longer be a piece of wood. So wood is not one atomic atom. How many of you understand that? And the sap that pulses through it is not wood either; it is a chemical.

So you think that the kingdom of heaven looks like the pyramid.[8] But the pyramid is a template to get you to understand the different levels of consciousness and energy and time, and in a more sophisticated mind to get

---

7 The third plane or level of visible light.
8 Ramtha's template. See Fig. B in the glossary.

you to understand what are the natures — what are the natures — of quantum particles. And clearly at this juncture — those of you who are bright enough to take the leap — everybody here understands that every one of those particles is alive. They are not simply dust in a dust storm; they are alive. They are cognizant beings. Hard to imagine. So as you try to shrink your large and heavy world down to the small, then the paradox is that how could life keep living there in its original form after what could you possibly estimate as time? When you think the kingdom is only this big, how could it be possible that there could be expansion in the kingdom? Well, therein lies your ignorance.

In this creation of life you never created the end, because in the mind of God there is no such thing as ending. And God's great effort at analogical mind instilled the breath of life in everything that was created. In these kingdoms, nothing has died, rather everything as a lifeform has evolved. Everything is evolving. It means it is changing and it is doing it accurately and perfectly. So there is no death there. And there is, you would say, "All right, are they eating each other?" No, they don't have to, because the idea of food wasn't an idea. It may well be sometime in their future. But consumption had nothing to do with keeping their body alive. It was rather consumption of an electrical field, and that happened to be their atmosphere.

What isn't God? If God, the great creator, the great gardener, is leaving fields of flower and fruit and nut, fields of animals, what is that that comes back to God? The mind of God, because this is the gift of life, and its activity called mind is what comes back to God. What is not God? Who is not God? What ingestible lifeform would not return to the mind of God? How can we presume such frightful human beings, living as if there is no tomorrow or threatening to terminate life as a torment and punishment to others around you, the ultimate pain delivered? How can you who live in such a struggle for life — which you do, which will become clear — could you absolutely understand that your mind is returning to God and that you shan't go back to the Void? That is impossible. That which God has created, indeed that which God has analogically given life to, belongs to God. And how do we dance in the mind of God? As individualistic and as lovely and as beautiful as the twinkle in God's eye, when God leaves that breath of life, for we consume God. And even ingestible bacterial forms of matter are God and were ingested to enliven. That was the purpose. And what is the reward? They will always live.

What Gives Subatomic Particles Their Weird, Dynamic Nature?

How many of you have been reading about quantum particle physics? Raise your hands. Really? I want to ask you a question: Is there any particle in the study of quantum physics that is the genesis particle that is all particles, or are they all unique and different? (Audience: All unique and different.) Now I am going to ask you a further question: Are those particles alive? Are those particles alive? Are they alive? Those of you who have studied quantum physics, tell me: Are the particles you are studying, from neutrinos to electrons to positrons to atomic structures to quarks, gravitons, are they alive? Then why do you read them as if they are not? Why do you read them as if they are not alive? Is a proton alive? How about an alpha particle? "Oh, they sure are." How many of you know what a graviton is, a graviton?

Father, what is the great argument in the church about how many angels there are on the head of a needle? Are the angels from a different realm? Then they shouldn't occupy any space at this realm. But what if they do, on the head of a needle? What is the argument: How many angels are on the head of a pin? But what did they ever conclude angels were? Do you know how many atoms can exist on the head of a pin? So what is the difference in an atom and an angel? (Answer: No difference.) You got it. So maybe you should write about that little argument.

Now so even though many of you don't emotionally want to agree that particles are life, then let me tell you where you are headed to: that you will not use your natural faculty because you don't believe in it and because it isn't the instant gratification you are looking for. But to many of you, you will stay glued to this and understand that perhaps in reading quantum physics, what gives particles their weird, dynamic, individualistic nature on the quantum backdrop — stay awake — why they can be in any one place in any damn time they choose to be, and so the measurement of where they will go in approximation to mathematical theory is a moot thing. Who is telling these particles to fold and unfold? It is because they are intelligent lifeforms. And does a graviton only exist as the subatomic particle, or can a graviton actually be traced all the way back to the origins of the first plane? What is a graviton? Graviton, Newton's gravity: A graviton is a multiplane particle. Why? Because the nature of a graviton is an energy that has been collapsed from the intent — from the intent — and the analogical experience of the God who gave it life.

The intent in the mind of God is reflected in a particle that is wild, that is difficult to ascertain with any other particles. They are called gravitons, and they are dimensionalized and energized, that what is apparent in the quantum field, that that graviton can be apparent in all of these fields and it exists at the exact nature in consciousness. In other words, a graviton is the glue that keeps the intent of God and God's life in it together. So it is the intent of God that it should remain, and we can break down God into these finite particles and give them justification. So what is the gravity of the biggest ball you know, the Earth, rotating in its gravitational field? Well, now we know that gravity is the intent of life to be sustained.

Now how many of you now understand the nature of a graviton? Raise your hands. So be it. Now are particles alive? And is gravity an intent? So do we see gravity simply as a rotating electromagnetic field, or is it a lifeform of intent? (Audience: Lifeform of intent.) If you were really an enlightened student, you would start having some immense realizations right now.

OUR QUANTUM STATE SIGNATURE

So we are going to begin with a statement that Shakespeare made. It is an easy one to remember. He said, "Ah, we are the stuff of dreams."[9] I would say that we are dreaming beings in a dream held together by the ultimate dreamer. Why do we refer to this then, indeed this knowledge, as the stuff of dreams? And why are you referred to as living, dreamed beings held together by a greater dreamer? Because in consciousness and energy — God and secondary consciousness — secondary consciousness, and wherever it inhabits, is the dreaming being. It in turn is made out of dreams. It in turn has the faculty to create dreams. And all the while that it is doing all of this, the one great dreamer, God, is now knowing itself, dreaming of itself, self-possessed of itself. And that is a beautiful statement, because when we get to study this area of amazing science we begin to dip into that which is termed quality of a world, particles, changing particles that don't travel but fluctuate. We are in a world that in this quantum field that nothing travels anywhere, because in the quantum field there isn't time. It is a field, indeed, in which that tiny field is supporting this larger picture. And in this field the Observer, either casually

---

9 "Our revels now are ended: these our actors,
As I foretold you, were all spirits, and
Are melted into air, into thin air:
And, like the baseless fabric of this vision
The cloud-capp'd towers, the gorgeous palaces,
The solemn temples, the great globe itself,
Yea, all which it inherit, shall dissolve,
And, like this insubstantial pageant faded,
Leave not a rack behind: We are such stuff
As dreams are made of, and our little life
Is rounded with a sleep."
William Shakespeare, "The Tempest," Act IV, Scene I, The Complete Works of William Shakespeare, Art-Type ed. (New York: Books, Inc.), p. 17.

or commonly or distinctly, is focusing, thinking, contemplating, reacting, is causing this field to blink out and reappear in total support of the dreaming mind of who you are.

Now having made that statement, let us make now a far-out statement that even Plato exclaimed. "All matter is alive," he says. "The world is alive. The universe is alive."[10] He is correct. You don't live — right now, I am talking to you — in your life you do not live in a life of objects, nor do you live in a world of objects, but rather you are living in a life, your life. All those objects are called experiences. You are living with experiences, not objects, and that the world, indeed that which is termed the universe and universes, space itself, is not about grand nebulae or dying suns and reborn suns but rather it is not about the objects. It is the experiences, so trade objects, things, for experience.

Now the way you think, you think in terms of things: people, places, things, times, and events; correct? So you wouldn't see, for example, anything other than an object in which you could use, as if the object itself was composed of unlife, particles — that the very matter that it is composed of, the very atoms that give birth to the grouping to make an object its distinctive, individualistic self — you don't think that the particles, the dynamics of particles, atoms being birthed off of the subatomic field, you don't think that those particles — you think they are like dust that are somehow held together with compression and pressure and atmosphere. So you are thinking then that all objects are essentially inanimate, they are indeed not living things. But if you continued that vein of thought, then it would be impossible for you to interact consciously with a dynamic field called quantum field. It would be impossible for you

---

10 ". . . nor did any of them qualify at all for the names we now use to name them, names like *fire, water,* etc. All these things, rather, the god first gave order to, and then out of them he proceeded to construct this universe, a single living thing that contains within itself all living things, mortal or immortal." Plato, "Timaeus," *Complete Works* (Cambridge: Hackett Publishing Company, 1997), p. 1270.

to interact with it and to change it because in that field, all that is, all objects, are experiences because they are comprised of, indeed as we have studied, the stuff of analogical consciousness and energy.

Now let's refer back for a moment to the concept of analogical. Do you recall what analogical is? What is the purpose of analogical consciousness is that the idea can be given consciousness in that analogical experience in which Point Zero and primary consciousness becomes the object, becomes it. When that is finished, when these pull apart, it has in it the breath of life, which means it has in it consciousness and of its experiences now will develop its own singular mind. So now if we think of it in that term, then a ball on the seventh plane might then begin to appear as one of those virtual particles flashing like stars over the moonscape of the quantum field. If this simple object was created out of analogical experience from God, if that single object now is an analogical experience of God, then it follows that all things in this snowfall, as they move deeper towards Point Zero, would not then be a simple electron. It would not simply be a simple neutron. It would not simply be a quark, a fermion, a boson. It would not simply be these sterile particles that you are talking about, that in fact if you could see this dynamic field, remembering all the levels, and if you could peer at all of these levels and see the dynamic beauty of this field, you would understand that all particles are consciousness and they have a mind and they have been experienced, so they are experience.

So when you were endeavoring to understand the nature of a fermion or a boson, those are names associated with the scientists who discovered them so they are being called by their discoverer their own name, his own name. But in reality those particles that you were endeavoring to understand how their relationship works together, perhaps now you do, but we cannot look at either one of those groups and simply say that this is a handful of particles over here and another hand over here and so we put them all together and we can make a sand ball — we cannot think that way — that all the particles you were studying on are alive. They are alive. Not only are they alive, they have multiple unlimited ability to be in any state that is therefore equal to the mind of God, and we now know how extensive the mind of God is. Furthermore, while you were studying them, you were interacting with them. What, do you think you were talking about them behind their backs?

Now all of those particles then form group particles, but they only form that group particle not to be random but to follow exactly your mind. Now I want you to write this down: The quantum field indeed is inextricably combined with mind, and the actions of mind is the Observer. So now if the quantum field is inextricably combined with mind, then what levels of mind is your particular quantum state aligned with? First and foremost, God is what keeps it together. You — you, out here — however and whatever you do, think, cause thinking, cause reaction here, your quantum state is instantaneously maneuvering, is maneuvering and adjusting you. Everything you see is objects in your reality. Everything is adjusted on a quantum scale to what you are thinking. This quantum state belongs to you.

Now you all have to understand that what you are learning here at this level is the nature of your mind, not your body, not your emotions, not the things in your life yet, not what you are wearing today, not what your face looks like. We have isolated everything to feed and to focus upon your mind and its nature. Today we will understand how the brain can process the quantum field, can process it into life. But for now, this is a very selfish talk because it excludes what you want. It excludes your comfort. It excludes your body. It excludes your dreams. It excludes everything. And we are isolating down a group of knowledge, imperishable, that you have never really focused on before, food for your mind, and understanding how your mind can possibly interact with matter.

So this teaching is to the glory of God, the Observer, and the mind of God. And until we can address it and make room to understand it, then all other teachings that follow after today are going to be laden with doubt and question because you didn't take the moments to isolate your mind and contemplate its beautiful, wondrous, and marvelous capacity. So now at this point I am going to talk about that mind, that Observer.

You, any of you, have made your descent and unfolded here.[11] Each one of these levels your mind is connected to, but new body, different progress in mind, different quantum states, different consciousness, uniquely. Next level: new body, new mind, new particles, new quantum field. How many of you are understanding? And it is distinct. How many of you understand that? So be it.

So now you are down here. So what was it that was able to descend or unfold into this very, very slow and large reality? Was it your body? Was it your emotions? How about the color of your hair? What has survived as a transient through all realms and all bodies and all distinct levels of consciousness? (Audience: Mind.) That is right.

Now doesn't it make logic to you — speaking to your mind, that is, of course — that a consistency of descent towards individual exploration, individual adventure of making known the unknown, dreaming dreams, wouldn't it make logic to you that there would be a consistency of descent from you who finally unfold into this plane? And wouldn't that consistency be that not only is your mind now here functioning on this river of consciousness here, this river of energy here, this time, this distance, this space, but that that mind has survived and been the very element, the very divine process that has now started the body and its reincarnation, its working mechanics, to work and labor in this consciousness, to create in this consciousness analogical reality that becomes a living state; otherwise

---

11 The journey of involution through seven levels of consciousness and energy from Point Zero to physical mass.

how can you say that you just are a new soul that just was born? It is ridiculous. How can anyone say you are new or old? Anyone who says that is ignorant of the kingdom of heaven and the process of life.

We unfold in every one of these realms and planes to come here with an individual signature, and that individual signature is our mind. And we have brought the capacity of that with us, as we are about to learn. Now I am not speaking yet of your thinking brain. I am not speaking yet casually of the memories you have stored in that brain. I am speaking of the descending mind of primary and secondary consciousness, secondary doing the descent, that will always be you. No matter what body you are in, no matter if you die and you are caught back up in IR,[12] it will always be you. You are never lost to God, and you can never be destroyed. So what composes the descending mind? God, secondary consciousness, the achievements of life taken from each level, each river of consciousness that life is now given. And as the expansion occurs, the life is held together because of the mind of the descending entity.

So how is it then do we think of our mind? Do we think of our mind in terms of what we are thinking today? That is not our mind. Do we think in terms of our mind in how we look today? Is that our mind? How do you know? The mind then is as invisible as that quantum field, and yet it is the mind that brings the quantum field into mass. So your mind, as we speak this moment, has prepared a quantum field, a quantum state in which you, your body, your brain, is being held together in the quantum field that created those mysterious things called fermions and bosons. You are being held together as a large object for the vaporous thing called mind. Moreover, your large bodies, your large objects, your life — your life as we now know — your life is equal to the objects that are now experiences of people, places, things, times, and events that are your life because you created them and experienced them. And that is the

12 The Infrared.

term of your life. Mind is those achievements. It is mind that fluctuates in what is called a quantum fluctuation in a quantum field, and what that means: a shift from one state of the field into another state of the field.

It is mind, therefore, that is quantum fields, quantum particles, life that belongs to you, experiences. You are making new realities out of particles of wisdom. You are using the same particles for when you were very tiny now to create the very large in your life. So your mind, if this is your journey — any one of you; it doesn't matter — if this is your journey down here, then all the quantum particles in these fields are your experiences in descent.

So now you made this descent to get here. All right. And all that you created, you and your God, because the entity that is sitting here in front of me is secondary consciousness fighting to be recognized in an emotional body. So you all made this journey. Are we to assume that you got here without creating the reality, the life, the platform in which to exist here? Do you think you just bubbled up from Point Zero and floated down here? How can we then begin to understand quantum physics and the quantum field unless we understand our interaction, our absolute intrinsic interaction with all these particles; otherwise who can explain why observation in a particle field is so difficult? Who has ever tried to measure the velocity of one particle appearing and then disappearing and then trying to put a ruler to it and measure the velocity of where it appeared, disappeared, and reappeared when the quantum field becomes the ruler? It can't be done.

Now so how close are the scientists who study these particles? They are actually studying their own experiences, and they are intrinsically combined. So all of you have created what we call quantum states, and the quantum state is exactly where Mr. Bohm's implicate and explicate order is. All of you have a specific quantum state. It is your signature. It is your signature. Now so you see all of these particles? Remember we told you that they were alive?

Remember how you learned that they got there, through analogical experience? How many of you remember that? So they are alive. So who is the creator of your state? You are, because your very large, big body in this very large, enormous, immense consciousness came from all those particles and is governed by all those particles. No thing large or small exists without life.

Now so then the mind that has encountered and created on every one of these levels, in a rite of passage, got to unfold here. So then if we looked at that field strictly from a side view, because it gives character and landscape to it, so your field would look something like this.[13]

FIG. 1: OUR QUANTUM STATE SIGNATURE

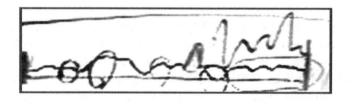

Now every one of these dips and valleys here — for example, this, this, and this — the quantum field is made up and it looks like this, as a skeleton. And what these are, these heights and depths and valleys in this landscape on the quantum field, are the formulation of particles into mass, that those particles then, this becomes the pattern of not only how your body is made, what the DNA will be made up of and the potential in that body of how your brain will think, how it will construct reality — along with that everything in your life, from the wood you have made your hovels out of, to the automachine, to the gas you put in the automachine, to the clothes on your back, to everything, every grain of sand that you call your life — all those objects have to follow this particular quantum state.

13 See Fig. 1: *Our Quantum State Signature.*

Now it doesn't look like much but then again you haven't seen the quantum field either, so this is an example. But although it is simply but mysteriously drawn with spirals and valleys and elongated plateaus, it is exactly what a quantum state looks like according to the mind-set of the Observer. Now why is this so important? Because it is this quantum state that is also keyed by the soul, so information is keyed into this state by the soul. And so then what does this state mean? This state is an unfolding state from childhood, from birth to death. Now why was that keyed in? Is that preordained destiny and, if so, where is free will? could be the argument. But one could only postulate, as it were, that argument if they didn't have the rest of the story and didn't understand science in its most mindful matter.

Now this state then means that in the world of the tiny, this is the pattern that you are created from and the engines of your DNA is created from, and every cell holds and is this quantum state. So we can't see this state as something that is out there. We can't do that. We have to subsume this state as us, like the stalk of the green stem of a flower that is pushing the flower through the green tube. This is us. I have a state; you have a state. Why then is this the pattern of your destiny? Because in order to have a body, it must be comprised of intent, and the intent of the body is for the experience of finishing what is yet undone and then making known the unknown, the engines of creation, as it were. So indeed your mind, your Spirit, your soul, your secondary consciousness is following and is sprung from this pattern right here that is now going to engage this long thick river of consciousness. How many of you understand?

Now let's go back for a moment. All the large things you see around you are alive. But did you think that these things were flung out intact at the big bang? Are there refrigerators still floating out there in space somewhere? No, it is coagulating matter. It is coagulating matter. So then everything large around you is actually comprised of matter, and all matter is comprised of atoms, and all atoms are created from the subatomic quantum field. So is there a disconnection from the large to the tiny? Is it the large that makes the tiny?

So I want to ask you a question: Do you still believe that the large is large and the tiny is tiny, or do you see a correlation? That is wonderful because if you would have said no, then you wouldn't have been as sophisticated and as enlightened as those quantum physicists who know exactly that all large comes from this dynamic field. That is why they are on it, and that is why they are passionate about it. That is why they love it. They are the greatest mystics that live, are quantum physicists.

So now then it would seem, as it were, that then as you grow, then in your reality your body, your life, people, places, things, times, and events would have to follow this formula.[14] How many of you understand? So, in other words, nothing outside, nothing below, nothing beside can change the way you are creating reality because you are creating it from a static quantum state.

There Is No Time in the Quantum World

Now let's talk about why this state is the same state that you were born into this life with and still have. Let's talk about that. If you are working in time and your body is changing — time, you know the story; you are getting older, things are wearing down, or you are young and stupid and everything is stupid — well, then you would say, "But, Master, I have changed. My body has changed. I am not what I was when I was six years old. Indeed I am not even that which I was when I was fifteen years old."

And I would say, "Yes, you are, because you are absolutely following your quantum state of mind."

"Damn."

---

14 Our quantum state or signature.

"Yes, you are, because here you are at this age, and this age, and this age, and this age, so in fact nothing even at this age was not destined to come in your life. Your destiny was to live this entire quantum field, and nothing new, nothing different is going to happen to you because you are following a meaningful pattern."

Now when we talk in terms of states — Now this is interesting, so pay very close attention and then you will understand how dazzling the kingdom of heaven is. This is a state.[15] From birth to death, this is our state of being. And nothing in our life can enter our life — we can't manifest anything in our life — that isn't already in this state. How many of you understand? Now the reason it is called a state is because this isn't about time. There is no time in this state. You know, as the Good Book says, I am alpha and omega. In the beginning, there was the end. This is that referral. How many of you understand? So now in the quantum world, in the quantum field, things don't move. If they did, then there would be time in the quantum world, but there is no time in the quantum world. Now so things don't move here. What they do is fluctuate and reposition the field, but they disappear. They enfold and then unfold, but they don't move. They don't move. There is no time in a quantum state; therefore that is why it is called a state. And it is a perpetual state in the quantum field. There is no time there. All that you have been living you have already, in your state, had. So it is called a state because it has no time, because your birth is known and so is your death.

On the face of what I have just told you seems such a blatant contradiction to what I have been teaching you, doesn't it? No, no, let me explain to you in these simple terms. In a quantum state, all past, all present, all future exists simultaneously in what we call the Now. Now there is only one place that the past, present, and future are in a constant Now and that is in the mind of God. And indeed

15 See Fig. 1: *Our Quantum State Signature.*

the mind of God, as it were, is what comprises the quantum state, which is your individual state. In other words, this is secondary consciousness.

Now so all things are already known about you. So then you would say, well, why bother learning anything if my state of affairs will always be? Well, because all particles — remember those little alive particles, consciousness and energy, experiences — all of those exist simultaneously in all levels, in all potentials, and in all possibilities. So what would that mean? So would that mean then that the very same particle quantum makeup of my state can be the same state but in infinite states of possibility? Yes, and all we have to do is maybe change this state right here to something that looks more like this.

FIG. 2: QUANTUM FLUCTUATION

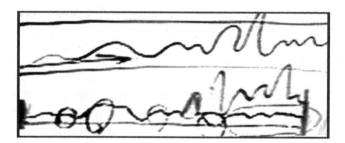

Now notice I drew over the former state, so the former state is still intact, isn't it? But is it changed? How many of you agree?

Now so your state and all that it is comprised of, every position of every intelligent particle, every experiential intelligence of every grouping and its consciousness that is affecting mass — Remember, remember, this effect trickles all the way down to the dirt in your garden. It isn't just "My state has changed and here is nature." You cannot have a quantum state change unless that change is consistent throughout people, places, things, times, and events. Now if this then is the static state of your quantum field, your quantum state, that has given you life — is everything, is the pattern in which all things can manifest and cannot manifest in your life — if that state then can be altered, then is it really an alteration of the state or is it a possible state that exists already?

Are you jazzed? How many of you are learning? So now in the immensity of space, the immensity of an Oort cloud that dwarfs the Milky Way, in the immensity of all space, the very basic chemistry is a derivative of this quantum field and those particles that you have been studying. So the same particles that make you up and indeed are the basis for the things in your life — and here are how many possibilities an atom can become — are also the same particles in planetoids in the Oort cloud.

Now that didn't hit you yet but the very same common particles — You understand what they are; right? We are not talking about exotic particles here; we are talking about basic particles, yes? Those basic particles indeed have collapsed according to this state to be a hallway in which you view your life, and everything must conform to it. Agree? Now the same particles exist in planetoids in the Oort cloud that comprise those planets, those atmospheres, those beings, those gases. Those exotic places are still built from the same basic particles. And yet the possibilities of the one you are learning about here have the potential simultaneously to be unfolded in the Oort cloud and planetoids, the same particles, the same ones according to a quantum state.

So is it possible that from your quantum state, and the atomic structures designed from it that constitute your mass, is it possible that all the structures of those atoms and indeed those subatomic particles are also manifested in the same moment in the Oort cloud? Yes. But I just told you that was your own individual state, that was your own quantum state. Yes, it is. Yes, it is. But in the quantum world with no time, we would only find this irregular and discontinuous if we insisted on seeing the quantum state as a state of time. That is when it would become irregular and discontinuous, but it is a state. We call it that because it isn't about time. So that which exists as a static place is called a state; it is not changing. How many of you understand?

Now this state is past, present, and future, all in the same state. Moreover, it is comprised in the quantum field. Remember the mind of God? So all your quantum states don't change; they fluctuate. And it is called a quantum fluctuation. Now you got that? So then this is you. This state is composing the matter, the large objects in a state that you exist in. Indeed that state is also very involved deeply into the quantum field, as we will get to today as well. However, for this moment, a quantum fluctuation would have meant not that there was a change in this state, but a state that already exists as a potential now has unfolded as a new state. Remember enfold and unfold — how many of you recall that — of the implicate order? So now this state is now this state, or this state is now this state. Maybe in actuality it is the same state except shifted into a possibility that is now a state. Which do you think is the proper answer? Which do you think is the proper answer? Hey, don't you want to know back there? What did it do? Because, you know, whatever you answer is how you are going to work that state of yours.

Parallel Lifetimes and Quantum Fluctuations

So now parallel realities, parallel lifetimes — Listen to me. Turn that beautiful brain of yours on. Listen to me. With what we have thus learned about that which is termed constant states, constant quantum field state — got that? — now in what we have learned, because it is a state and it does not process the past or the present or the future as anything but what the state is, so there is no time, indeed then if we are talking about parallel states we would be talking about parallel experience possibilities.

Are you learning? Beautiful. So what did I tell you about objects? Is it objects or what? Experiences. So what are particles? Are they particles or what? Experiences. That is the answer, because it is consistent with the nature of how they were created: analogical mind, God, and life. Right? Let's don't leave that out, shall we?

Now so then the nature of a parallel state, or shadow state, of the quantum field would be referring to that this present state, according to its quantum arrangement, its landscape, by altering or changing any of the landscape we have changed or repositioned the arrangement of experiential particles. So if we alter anything in this state, we have created a parallel state. Now so are there indeed shadow states that are yours? Are there indeed parallel states in which would form indeed your mind, because it would be inextricably combined to it and indeed would be the pattern in how your life would follow, what would be in your life and wouldn't be in your life? Are there alternate states that would suggest simultaneous lifetimes?

Now how many of you — come on, use your brains, use that gray matter up there — how many of you could understand then if those subatomic particles, as it were, that constitute indeed at the quantum level, that those particles don't live in time but one particle could be in infinite positions at the same time? How many of you could see that? How many of you could see that? So be it.

So now why is then that the awareness of your particular quantum field, if indeed it has these parallel experiences and indeed probabilities, why is it the static field that it is, without the mobility of becoming an alternate field? Now in reality, true reality, these alternate — alternate — conjunctions of this state are infinite. But we don't see them, as if we had all the boards in here around the wall we could draw infinite states all the way around the room. It isn't the way it works. The way that it works is that this state enfolds and unfolds a possible state within its state existent. So, in other words, this that we see here was an enfolding of an alternate lifetime or, at this level, an alternate state. Got that? So, in other words, the dynamics of the field is that this[16] doesn't blot out and then we move to this[17] field. We don't move anywhere. In the quantum field it is an unfolding, and that that unfolding creates indeed that which is termed the landscape of your personal quantum field or quantum state. So how many unfoldments can this one little state possibly become? Infinite. Just like how many people in the world have the same DNA, but they are all DNA — how many of you understand? — but even greater when we talk about the quantum field.

So now if all of you understand that the state itself disappears and reappears, and in its reappearance indeed we would have — if we didn't know better — we would think that we are now looking at a brand new quantum landscape. So we would say, "Ah, well, this quantum landscape is different than this one, and if we can photograph it, we can measure its influence and its signature difference from the one previous." Are you with me? But in reality it has unfolded from a prior state into a new state.

If your own personal quantum field, your own personal kingdom of heaven, your own personal field in static self that is making you who you are now, if that field were suddenly unfolded into what we could only call a parallel state, that now in the state here that these fluctuations within the quantum landscape would mean what to you in your life? How would your life be affected? Because, remember — now I want you to get over this — we are responsible for all experiences, and therefore the power of our creative mind goes right down to the soil in which we live on, so and all of that is alive. Now I want to ask you a question: If suddenly your shift went from this to this, how would the ground of your being and indeed the ground of your reality be? You may begin.

---

16 Our prior quantum state.
17 A new quantum state.

So now I love watching your bands. I love watching your brain producing mind, and you can tell it by spikes in the person's bands that they are engaging something they have never engaged before. And it is titillating.

Whenever you read about these terms in quantum physics — they can call them anything they want — that in understanding when you are reading about a simple quantum leap, then you are actually, you cannot help but be, reading about your own quantum field. So was there ever prophecy that there would come a book in which true gnosis — from the unbroken love of God to man and woman — would be revealed and that looking upon the pages manifests the kingdom? There has been. You are doing it now. You are doing it now.

Remember that this Assay means test, and the test is to stay focused, listening intently without distraction. That is the test. It is a greater test than going to the labyrinth. This is the test. So why would that focus be so beneficial to you as a test? Because the very nature of what we are talking about is the very nature of your mind — remember, your mind — your reality, your God, experience, and possibility. Now what were you just discussing? Parallel states. Now you didn't know that before, but now it came out of your mouth, came out of your mind, came out of your brain. Do you know what is going to happen now? You should be happy. You should be happy, every one of you.

So now you see I have been talking about you all day. See, you have been focusing on me, that is showing you "you" all day. So now if you then — and blessed be you that you can see how these shifts can occur within a static state. You can see that, yes? How many of you can see that? So it isn't — it isn't — then necessarily counting states, because if we did then that, we would be violating the law of states. We would be back into time. How many of you understand?

So then these are used as metaphors for people to follow who think in time but in reality is one state. Here it is, and that is yours. That is yours. That is everyone's.

That is everyone that has ever been, ever has been, ever will be in any dimension, on any Earth, in any body. That is it. The beauty is, is all that ever has been, all that ever will be, from everyone that ever was, is also in your field because they have modulated their state in a different way, and your state can then have a quantum fluctuation to be exactly that state.

Now so then we see that it isn't about parallel states, that in reality it is about how this state fluctuates. The heights, the valleys of this state, how they change is really — The most minute change is a shift in quantum state. Now so is it possible then that a lonely human, who starts out with this state, could then learn the most fantastic knowledge and in this state never die? (Audience: Yes.) Is that consistent with the DNA immortal gene? (Audience: Yes.) Is it? (Audience: Yes.) Is it consistent with that the brain that you have, you have only used ten percent of? Would those states be consistent with the unused portion of the brain? (Audience: Yes.) So now how many other synchronistic examples do I need to show you?

Now so in your state that you have now, infinite possibilities, and when we talk, as it were, about objects being experiences, not objects, then we are also talking about particles indeed being experiences rather than simply deadpan dust, but they are actually fighting and alive. Now that is a key here, because if then all the particles that comprise your quantum field are also existing in a past, a present, and a future possibility, at what length you cannot exhaust except still using the same quantum field, so your particles are alive. They are multidimensional; indeed they are multiunlimited. Their capacity towards experience in a grouping, no one has the math to infinity.

If we have a quantum fluctuation upon the landscape of this your quantum field, if we have this fluctuation, then the field has shifted from one's field into a possible field now occupying the same state. So that would mean a parallel field has just shifted into the constant state. What would that mean? Does this shift of parallel fields into one common state, how does it then from its shift affect your life that seems to be indeed very smooth, very continuous? There is the same house. There are the same flowers. There is the same thing. And I am telling you when I started this that this state here regulates all people, places, things, times, and events in this state. No thing can come into your life that doesn't share this same state.

Are you just about ready, through this knowledge, to depart from the concept that you are victimized by your reality? (Audience: Yes.) Are you about to depart from the concept that it is you against your life? (Audience: Yes.) Are you almost — almost — ready to do that? (Audience: Yes.) Are you almost ready to depart from the concept that there is you and then there is God? (Audience: Yes.) Good, because if you can follow this, speak it, know it, then you are right up there with the greatest and most brilliant minds in quantum physics who are delving into the deepest mysteries of its potential. You are learning them here. They are digging and we are digging. So now if you are ready for that departure, then you begin to understand something about yourself, that then why were you a victim of your life's circumstances all these years?

Now, masters, so do you really think that you are such an individual that you are not included in quantum physics? Do you think that you are so individualistic that somehow you, your quantum state and its resulting life, were a vast right-wing conspiracy from God? Are you about to depart from that you are so special, so troubled, so traumatized, so dulled by life's harshness that there must be some secular, individualistic punishment that the quantum field is dealing out to you and for some reason you don't know why? I have got to cover all this stuff before I move into the next section. Are you about to depart — now hold on; everything is purposeful — are you about to depart from that conservation of lack? (Audience: Yes.) Are you about to depart from, say, for example, being so damn happy you have everything you want but feeling so damn bad that others don't? Are you about to depart from control, that something out there is happening to poor you? Are you ready? No, because you continue on after I have taught you all this — you know, ignorance is forgivable but the abuse of knowledge — you deserve it.

So then that brings up this reference of that previous, earlier statement that mind is inextricably combined with the quantum field. And if it is, what is the controlling force that designates, out of this very large river of consciousness, the body? The way the body is put together is based on this. The environment the body is born into is based on this law. How the body will grow up will be based on this law. Everything that mind, indeed secondary consciousness, will experience is going to be based — are you listening to me? — in the ground in which all that will bloom, and it will bloom from this ground. Do you understand?

DECODING THE LANGUAGE OF THE SOUL

Now in the quantum world our signature is often controlled by an anomaly and phenomenon called the soul, and the soul is, of course, the ultimate recordkeeper. For example, if our mission could be as simple as what I taught you the first day, that indeed God in its eternalness that contemplated itself — and that contemplation is the key to expansion, even in quantum flux — that in contemplation, it created secondary consciousness and began the magical journey of selfishly experiencing and loving itself. And we are that self and indeed that journey. The soul kept a record of the projects we started, projects we started.

So now what the soul did was — Think of it as a book, a large book, with pages made out of light, and on those pages is a secret tongue, that the tongue on those pages is symbols. And maybe if we wanted to break the code, we could start perceiving these symbols as quantum fluctuations. So maybe what we think are symbols are really states of conscious fluctuation. Would you turn to your neighbor and tell them that. That is, by the way, written on a single page. It looks like hieroglyphs but it isn't, and it isn't the fire words of the Cabala but it is similar because each letter is saying something, that the only way we break that code is when we understand that the formations of these letters are formations of quantum potentials.

If you opened a book, read a mathematical equation, and then saw a drawing, if you connected all of the keys in the drawing you might come up with a template that looks like a bizarre figure. But if you laid it over the drawing — and there was the drawing of particles reacting in the quantum field, or fermions and bosons attracted to some, denied by others — and maybe if you looked at that and then read the description of what it is, perhaps that is what is in the Book of Life that no man can read, only the Holy Spirit can read.

Now this is deep occult knowledge. Make no mistake: The book exists, and no flesh and blood can read what it contains; only the Holy Spirit can read it. Well, now we understand what the Holy Spirit is: a quantum monad called mind. So let us come back now. So the soul then is endeavoring to configure, is endeavoring to configure this construction as a state that involved the configuration, the involution, which is moving down, that the book configured all these states as experiences, and in experiences those were given life. Those were the great engines of creation.

True and False Quantum Fluctuations

So the book has all of this information in it and it is in code, but it is in code in the terms of particles analogically given life and then built upon, magnanimous equations. And now we get down here,[18] and between that which is termed the light, the implicate and explicate order, down to here we have a problem. In the Book of Life, or the soul, you have unfinished creations that have gone beyond — beyond — simply fabrication. Now what if we decided to change, but we used our need as the basis of change? The need is climax. So we are changing from the old and lousy to the new improved. What is the central characteristic? The same climax.

Here is the key. The key is we are using the quantum state that is indeed comprising, constituting the affairs of our body and indeed all tissue. All life and how we interact with it in that reality is all part of our quantum state and that we are not changing, we are not creating quantum fluctuations, with every time we change a relationship. We are following the template by simply being that quantum

18 The physical plane.

field of repetitive climactic experience, and all we are doing is using mass to mass, our own quantum state in another, to have the climax. What could change? Now don't let your brain go to sleep, because what is important for you to understand is that in going mass to mass with any need does not constitute a quantum fluctuation of indeed your quantum state into a parallel state. So then how do we value change here? Then in your life, you have basically valued it in a sense that no matter what you do, you are compelled to create from need.

Now what does the soul have to do with that? Well, that is your signature in reincarnation, that in this quantum state that what you will always do in this state is have people, places, things, times, and events. I don't care what you do with them. You know, you can scale them from rich or poor, ugly, fat, thin, beautiful. It is all the same. You can move from this place to that place. You can eat this, different from that. You can do — you can search for all your needs. You are compelled to do so. Why? You can say, "Well, I moved out here. I sold everything I had, and I have made big changes in my life." No, you haven't. You have only operated within the law of this frozen state called your quantum field. You have only operated within it. New faces don't mean anything, because a new face in your life is a need climaxed. Do you agree or disagree? How many of you agree? And the tragedy is we know we are doing it. Isn't that so? So no matter what you do — you can get fabulously wealthy or fabulously poor; you can do anything within that state — you are not operating. It is not fluctuating. You are absolutely moving within that structure of that quantum field because no matter where you go, in the quantum field you are a state and a state of unfinished creation.

What does that have to do with you? How can a wee little old tiny state be such a driving force in your existence? Because it is the template of the soul that says you are going to come back here, you are going to be born from flesh and blood in an environment. You are blooming from this quantum state. Nothing — nothing — different will move into that state but all will obey its law, because in this state you have to keep meeting your climax until you realize what you are doing and have owned it. Only then do we have a quantum fluctuation. So now how many of you now understand that in your quantum state, that this quantum state was designed around meeting climaxes in people, places, things, times, and events? How many of you understand that now? That is the truth. That does not violate the law of mind that it is inextricably combined to this.

Now it goes — this is how deep this goes — that in this state, you can make all the money in the world. But in this state, that will not bring peace — do you understand? — because you have to keep repeating the climax. Do you understand? You can be the poorest person in this state, and in this state you are going to continuously meet lack. How many of you understand? It is a climax. You understand? In this state you can make anyone love you because anyone that comes into your field, you can buy them whatever they need, do with them whatever they need, furnish all their needs, be anything they want you to be. And how could they not love you in that lock-and-key relationship?

So even in these conditions you can have love but only under the restriction that every thing, every person, every place, every time, and event has to repeat an emotional climax. In this state you can be all to anybody. You can seduce anybody. You can get anything you want. And you just ask the hetaerae of this audience if that isn't so, because they have done it. You can be beautiful. You can be sexy. You can be there at their wanting wish. You can spit quarters out of your mouth. You can buy them anything you want, they want. You are there; you are there. You are like glue, and they will love you. Now what happens when they love you? Boring. You did it; you made them love you. To you, that is just another climax with no resolve, so you are going to get bored and then you are going to start picking them apart. You are going to start finding their flaws, pointing it out to them, reminding them everything was given to them, remind them you are the breadwinner, remind them that "No one will ever love you like I do."

So no wonder the moody blues comes upon you and you are so sad and depressed. And all your friends around you, they want to run and get you everything you love and everything you need, and they don't realize you don't love anything. You don't love anything. You are depressed because you are addicted, and you have subverted love in a calculating form for climax. What can they do for you? They are all in your quantum field. What are they going to do for you, bring you chicken soup? That chicken soup is now in your quantum field. Where can you go to be away from you? At every corner, there is an opportunity for a climax. What are you to do? There is even a climax associated with depression. There is a climax associated with sadness. There is a climax associated with everything you do, but what you have to understand, it is the same emotion only displayed differently in your life. It is the same emotion that is not owned.

Climax versus Wisdom

How many of you understand? Now in infrared, which is a reversal state, sort of an antimatter state, then in this state people who are earthbound and are caught up in their infrared bodies will attempt in this same state — which is always with them — then they will go and they will have climax, but they never get to climax. They will eat but will never be satisfied. They will lie but never be deceived. That is why there are entities that are earthbound here that are addicts. They are drug addicts. They are addicts, that in their vision they are tormented souls, and on that plane they are in eternal torment by what starts to bother them here. And there, then, there is no climax to torment as there is here. Here, torment has a climax called redemption. That is the high. In infrared, there is no redemption because there is no climax.

So then there is a whole place where there are orgies, people having sex in every conceivable degradated position like a pit of vipers all wound up within each other. Why do they stay there? Because every new infrared body that goes into the pit they are consuming, because no other association has brought climax. So it is called a tormented soul. The food that doesn't fill, the fire that doesn't warm, the habit, the game that is never won, the truth that is never found, the Holy Grail never found, all of that exists in infrared.

Now this is the great ancient wisdom about the quantum field and about that that field is inescapable from our mind. And what do we know about mind and how it manifests itself on this plane? That it is descended and indeed unfolded to, that mind then is that which gives the pattern of life for the quantum field for the person. The soul then designs that quantum field. The incoming secondary consciousness in body form is a living projection of that field, and their life of people, places, things, times, and events and environment are influenced by this field. And now we have a body, indeed a brain, indeed large objects, and we are now in the river of Hertzian consciousness and energy. Our brain — our brain — captures a frozen moment out of this eternal river. That frozen thought is in the brain, called a neurological thought. That thought then is built upon by more frozen moments out of the river of this consciousness, not the consciousness or energy of the fourth plane or third plane or fifth plane but the conscious river in which the buoyancy of mass has its domain. And so the brain is functioning in this consciousness, freezing thoughts into neurological forms, putting them into the frontal lobe — the frontal lobe — observed by the Observer, brought into manifestation in what is called a quantum fluctuation shift. And in that shift, the body simultaneously is prepared chemically, has the experience, the experience then towards climax. Then the secondary consciousness moves away from the climax back into the brain, and the total experience is now called mind, and life is now given on the plane of reality. God has planted a lovely flower. And as God moves on, behind God the plant is blooming.

So what are we to say about mind then? That then the mind then is inextricably combined with this static state, this quantum state, that the new life allows us to finish those concepts we started somewhere in this same consciousness, that in this life we are midstream in creation. We are addicted to the climax instead of the wisdom. And in this life our brain is going to add to the quality of the mind of manifesting those opportunities and either reinforcing the field as a static field, a frozen state or, in those interactions, come to a state of wisdom in which then the state is — disappears and reappears — is now a new state, because in the new state is going to formulate your new life.

Every attitude that you own — every climax that you need is an addiction, needs to be owned with the attitude behind it — when that is owned and brought back into mind as that which you have mastered and conquered, then the moment you have owned it this state reappears as the new state. And what does that mean? For your body and for all people, places, things, times, and events in your life, that is the corridor of projection from this state, that in this state — your life isn't changed, it isn't an evolution towards change, it isn't a slow realization — it is instantly changed in what we referred to earlier as parallel states, infinite possible states.

Well, in the large world here, in the bread-and-butter world of your life, what does a shift in a quantum fluctuation in your quantum field, what are the meanings of that for the life that you had previously? The moment there is a quantum shift in that field, you have shifted — you are not

73

evolving into — you have immediately shifted into a new basis for your life. Immediately you are now living a parallel lifetime, that that parallel lifetime is the large equity of the small definition of lifetime in the large and in the tiny-state shift, a shift in quantum state. A shift in quantum state brings a parallel lifetime and that that lifetime then, now everything is different. The relationship to you and your environment is lifted, for what compelled you before is not a compelling influence in the new shift. You are now in a parallel existence. In the parallel existence our mind does not leave our body behind in the old state but rather that the body that can also live in parallel existences because it is made of quantum material, is now shifted into the new, the new hall, the new life, and everything is different. What becomes apparent is that there is an at-rest of a climax that once before governed your life, is now nonapparent in the new life, and its influences are not seen in people, places, things, times, and events. That is the truth.

The keys to the kingdom of heaven is this knowledge.

WHAT IS THEN THE MEANING OF LIFE?

Now how many of you understand what a quantum state is? It is a fixed state. How many of you agree with that? How many of you agree that a quantum fixed state can be literally the life? And remember what we talked about way back there in the beginning about analogical experiences are now life? And I ask you now, what is the meaning of life? Well, now we understand that your life has been given experience to, and that that life is a projection of the fundamental tiny that already knows from birth to death if you follow — if you follow — and live by what you have not owned, you have an inevitable end. And that is your state. Change within this state does not create a quantum fluctuation. A mind-altering change creates a quantum fluctuation.

So then perhaps we begin to understand why there can be parallel quantum states of infinite potentiality. From that statement we go now to the large, because that tiny statement is responsible for the large global view of parallel lifetimes. Parallel lifetimes are equal to quantum shifts in states of consciousness, states of quantum field. It is the same thing. One is tiny, that governs the large. So now you are interested in a parallel lifetime. Are there parallel lifetimes here this moment as I am speaking to you and you are talking not about anything abstract but you are talking about yourself? When I ask you to explain, you are explaining yourself. We are not explaining an abstract here. So when we talk about parallel lifetimes, we are talking about quantum shifts of quantum states.

Now you are only interested in a lifetime. All right. So is it possible that as we speak, indeed in the same body you have, that it is possible without growing older to shift quantum parallel lifetimes into current lifetime? Is it? According to science, is that a probability? Yes. So this isn't New Age whales-in-space stuff now, is it? This is a possibility. So I want to ask you something: In your parallel life, the same body — the body doesn't have to die, doesn't have to go anywhere — the same you, what would your life be like if all your emotional needs had been met and satisfied? Think for a moment. Who would be sitting here in my audience in front of me? What if one emotion called resentment and anger — they are inextricably combined — what if that were owned, who would be sitting here today?

Let me give you another example. If we now understand that this quantum state is a soulful pattern of how — not only our life — how we think, what is in our life, our capacity to move within that life to infinite signatures of sameness yet never fluctuating the state once, if we know how that works then what would your life be like — listen to me carefully — if you were born into two parents who were so involved in their own emotional stuff they abandoned you, wouldn't you have needed that as a start in life? And why? Because your quest for love and being nurtured was the exact controlling power in which you would control everyone else by giving them everything so they would love you, but really in control, that the ultimate climax is controlling love. Then if we know that, then what would our life be like if we knew that that was the reason we had this life and that our climaxes after that were set up from the initial start to abandon us, to control love for the rest of our life? What would you be like today if you could own that? Would you be in a different life? Would the dynamics of the state have changed? Yes, because now owning that means that we have a quantum fluctuation in the state, and when we have a quantum fluctuation in the state, in the next moment the new lifetime appears, though it seems smooth. But what will not be in this lifetime is controlling love. How different would you be today if you didn't do that? How different would your life be? What would not be in your life and what would? All right, another example. Why are you insecure? Why did you create the emotional need to have the climax of feeling unworthy? It is all over your quantum state. Well, you know what unworthiness is? Let me give you a little hint. Remember today when I came in this audience and I said, "How many of you know more," and before I could get the next words out, you were jumping up and screaming and laughing and full of glee and cheerfulness? How many of you recall that?

Now what would we say about that, indeed, in terms of mindful reality? We would say this: that finally being compelled to enter the world of the very tiny and having to study it in order to come back to my school brought with you a brilliant and pleasant surprise. You understand now, with knowledge, about a science that is so elusive and unglamorous to the mob. You dealt in and now understand something about the quantum field that you never knew before, and that knowledge makes you feel so wonderful. Why? Because it means an opportunity. Why? Because it closes the climax of insecurity and brings wisdom. Knowledge ends insecurity. How different will you be now that you know?

Thank you. Thank you. It is your life. We have been talking about you, you know. It is about you we are talking about. It is about you, your life, we are talking about. Your standing ovation I totally accept and am honored by, but it has more to say about the applause and the joy and the appreciation about knowing about the deep mystery of you and your life. That is what the applause should be about.

How wise you must be already, how shockingly free, how understanding, you must now know. No gaps on the road to the kingdom of God. And are we toasting to a flight of mythical fancy or indeed are we toasting to pure quantum mystery, science, the Observer affecting the quantum field? This toast is to you, to your life, who hold the keys to opening the doors to changing states of quantum reality. So be it.

Breaking the Magnetism of Predictability

Now we move to a different phase. Now that we understand this constant quantum state — quantum fluctuations are unfolding infinite possibilities within that state — we could reflect that into the large world as parallel lifetimes in which we shift into with every meaningful ownership within our life. We are shifting — not changing — we are shifting into a new quantum state and a new large life. So now we bring that up to your large speed. And we can say here that emphatically — look at me, look at me — we can say emphatically now that you, your mind, your attitudes, are in control of your quantum state and are affecting your entire life. In fact, your life is built on your attitudes. Would you agree with that? How many of you would agree? All right. You accept that? In the old days they used to call that taking your power back.

So now how many of you agree that you are addicted to climax, whether in a sexual form, whether in a consumption form, any point that works up emotion to a point of climax? How many of you keep repeating the process? Raise your hands. God, I love you. Hold on, keep your hands up. Did you know that in order to be ennobled that first you must be humbled and agree and understand? Thank you for your honesty. Now hold those hands up again. Do you understand why the List[19] has not worked in your life? How many of you agree? Do you understand now how when you change mass to mass it does not bring wisdom? How many of you understand that? You do? So how many of you are controlled by your emotional addictions? I am bewildered and uplifted. That is the truth. And that belongs to you.

---

19 The List is the discipline taught by Ramtha where the student gets to write a list of items they desire to know and experience and then learn to focus on it in an analogical state of consciousness.

So now how does God create anything out of nothing? Do you remember the process? How many of you remember? Now the repeating of working up to a chemical explosion is now being done by your body and giving you that high. And then you come down and then you start to work it again, because that is what your state looks like. And nobody in your life — listen, nobody in your life, except someone who doesn't want to be in your life anymore — anyone who is wise in your life knows this about you and knows that the only possible reflection that you have, or they have to you, is if they can accelerate that need; otherwise they don't see anything else about you. So you are sort of stuck in this quantum state because you are working within this state to climax. Nothing happening out — There is no fluctuation in this state. Do you see that?

So now would you say then that your addictions started at the first climax? Now, remember, I am talking about climax from the most obvious term, sexual climax, or to children, crime and punishment, good and bad, punishment parameters, that everything we do has a climax, whether what we do is a climax of punishment, a climax of being good, a sexual climax that we woo and we lie and we do all the things just to have that climax. But the sexual climax, though the most obvious of climaxes, also infers other subtle climaxes like the being-bad climax or the depression climax or the control climax or the redemption climax: "We have too much peace in our life. I need to create chaos so that we can have romance." Romance does not come from peace. It comes from naughtiness, redemption, bonding. So if we want bonding, we have to create this whole route to create romance and bonding. Stealing, in this particular state, our goods and services — our goods and services for our needs — means that we must approach them within our state and with that which can come into our state in which we have to confront it mass to mass. So even our sustenance, however large or however small, comes from this state.

If we are clever liars, if we are the people who can make anyone fall in love with us, we are likely to be the most successful at getting what we want in life because these people can be anything to anybody, and the reward is to getting from them what they want. It means nothing to lie. It means nothing. It means nothing. All of it is to a climax of goods and services. How many of you understand that? How many of you don't understand that?

So even our need to be taken care of by someone — our need to be taken care of by someone — is because in this incarnation we have to be protected against ourself. So we must create someone in our life that will always be there for us because we know we are going to need these climaxes. But we need somebody to be grounded with, that will forgive us and nurture us, cleanse us so that we can go out and do it again. How many of you understand that?

There are those in this state that play that part really well. You are the constant. It is also your way to getting people to love you and to control people. You will be the constant in their life and they can go out and they can have these climaxes, they can do all these things, but you are going to be the constant player in which they all come home. When they come home to you, what you get to do is to control the methodology in which they are going to live, and it is called the climax of power. You are the one who loves. They leave home, prodigal son. They go out and do all of this stuff. They spend your fortune. But that isn't the climax. The climax is that you are the redeemer. You are the one who gives compassion. You are the one who forgives. You are the one who accepts the prodigal son home again.

The evolution of the state of your quantum state should become very dangerous, because in the old days — you know, in the old previous lives — you know, it was that rake and that whore or that countess, you know, meeting behind closed doors and having nookie-nookie and then showing up at the same ball with their partners, who they think that they have just expanded the relationship between them, not realizing that the partners had nookie-nookie at teatime. Now so you have emotions carried over, you know, unfinished realities from many lifetimes because you are addicted to the climax. And so every new incarnation is based upon a static quantum state that, in the times and consciousness now being known or made known or mindful of this state, then adjusts itself to the tremors and consequences. For example, in those days if you didn't have power you more than likely didn't have land, but if you did have land then it belonged to the princess or nobleman who owned it. So most of your crops were actually made for them, and you had scraps in which your family, as sharecroppers, could live upon. So if you were caught stealing or keeping more back than the landowner had, you had your hands chopped off. It was quick and easy justice. But the implications of that were catastrophic, because then who works the land? So now the children and the wife have to work the land, and the man has no hands. Are you watching this? So then he is redeemed by his family who now must be the breadwinner. And not only do they carry the load of their existence, they have to give a part of their existence in order to survive.

Today in this incarnation, this rapidly evolving, collapsing society, hypocrites are jeered at. Hypocrites are jeered at. Truth in its most carnal form is popular, isn't it? There is no sexual thought, there is no thought of reprisal, there is no attitude towards the rich, the poor, the black, the white, the lawgivers, the lawbreakers, that isn't out there in music form. It is popular to call people on their lies, isn't it? So today you cannot have a tête-à-tête teatime. You know why? Because so adamant is this consciousness on calling you a hypocrite that the truth reveals itself straightaway, because that is the nature of this Now mind. Furthermore, you don't get your hands cut off in this society any longer. What you have learned to do is to use people through political policy, through friendship, or making them trust you, and getting them to take care of your needs without you breaking your back in the sun bent over in the field, without you standing on the firing line, without you having to work hard for every dime. Someone else will do it for you, and all you have to do is just make them love you.

The truth teachings we had, remember those? They are the most hated and loved part of my teachings. Why was that so important to expose people? Because unless that can be addressed and shown to be what they are, it is what they will always be addicted to, and they will not be a quantum fluctuation in this lifetime because they will be owned by their secret. How many of you understand? How important is it to go into hell? Well, I was a cursing — I learned it well — I was a cursing, enigmatic God. But what was I doing? I was becoming analogical with the very people in which I sought judgment on and that that judgment would reach into a deeper order of their quantum state. Can't be — can't be — a God wearing wings and a white robe and condemn somebody because they have got a secret. You have to analogically be that person, and if you are, then by grace and association truth emerges and now we have fluctuations in the quantum field.

So, masters, how many of you have gotten a better understanding of the frozen state of your quantum field? How many of you understand that you have worked within that field without changing it? So be it. So now you know why there is a beginning and an end and it is already known.

In this quantum state that is unmovable, the biological clock of the body is ticking away. And the cost in which this life that was given to us by analogical experience — the mind of God gave us this life in the large — and this life is as those dream beings, in which a greater being is dreaming us. Then we are acting out in this dream either as a predictable dream or an innovative individual.

The quantum field and your quantum state suggest — according to you raising your hand and admitting that to me but, more importantly, to yourself — that is who you have to be true to. I am the teacher that asks the questions. You have to respond to yourself. Then if you do not retire any of these, conquer any of these to self-independence, to self-sovereignty — that the urges that you put upon and take away from people, your needs having to be met — if you can meet your needs by understanding it with this truth, the only way you are not going to die of your emotional affliction is by stepping on the stairsteps to heaven which, in this case, would be parallel lifetimes coming in at that moment and that your life is being shifted into new gradients from quantum fluctuation in your quantum field because your mind, indeed inextricably combined with it, has now been changed and expanded. The mind we started with is now evolved. The wisdom of those climaxes has ceased. The understanding of life that is given to them now is allowed without our interruption and aberration, and now we are no longer drawn to do it any longer. It has a life of its own. That is called detachment.

Every one that we do, we change our lifetime. We move into a parallel existence because our quantum state has changed. Every parallel existence will be an existence of matter that will not have within it the cause and effect of that initial emotional climax. In other words, all that is in it will not have that magnetism, that need. It will be completely different. So the existence from this ground spring of quantum thought will be immeasurably changed in its fluctuations; therefore the people, places, things, times, and events, the particles that comprise them, in your life will not be there as that which feeds a revolving-door climax but will be changed because the need is now wisdom, and so the magnetic dynamics in that has been altered.

And now we are starting to walk as wise people. And now we are starting to walk in a life we can see so clearly we have created by our changes, and that the more that we ratify those emotions into wisdom, the more profound the parallel lifetimes — and indeed the quantum states that are shifting — become. And it is possible — it is possible — in one lifetime to have understood this science. Remember,

remember, what I said to you: The very fact that you were studying bosons and fermions, you were affecting yourself by what you were studying. Either you knew it in deep degrees or shallow repetition; however, it was about yourself and about your life.

Reverse-Firing the Brain As the Key

Now when we understand that we are creating this reality and that there is a process in which we must obey, as the dreamt beings — as secondary consciousness in the mind of God as the ultimate dreamer — when we understand, we have to follow a process, that that process is the idea. The process is the formulation of the idea in holographic view, or energetic view, and then its ultimate analogical experience by being it, which we call the climax. But once the climax is done, we have pulled apart and seen that it is a living thing, that it is vital of its own energy, indeed of its own consciousness, of its own capacity to regenerate itself. How often have we selfishly, selfishly, created in our life the climax and did not permit what we created to regenerate itself or think for itself or be free of us? How many times did we have to control it and keep it in a cage so we could keep climaxing it? We have much to answer for.

It is possible, and now we have a glimmer of light, to see how such enlightenment, such tremendous knowledge, so gained, so grounded, is a neuronet in our brain. Well, if it is firing answers to islands of attitude, then it is the knowledge that has created the bridge in which the islands are no longer acting on their own but come back into one mass of experience.

It is possible in one lifetime, people, that such an aware student of such great profound knowledge would say, "I understand; where are the keys of discipline?" to be able to start making those shifts and be able to know immediately how reality has shifted, how different the interaction is from a former state now into the new state. But can we give up control of the former state? We have to. The former state is consistent with owning a wisdom, owning an emotion, letting it go, seeing the value of what it is and never repeating it again, never allowing that neuronet to fire independently as the source of manifested destiny. Moreover, this is where we separate the wheat from the chaff.

Those who are given this knowledge in such a profound display of science, who have the passion to experience truly the unknown, need not disappear or to be vaporized, need only to have engaged the knowledge into a rehearsed discipline in which thinking is rehearsed into rote, and that that rote then is the new state that has now had a quantum fluctuation in it. And now we have the new life, and now we are experiencing people, places, things, times, and events without need, without conquest, without climax. We are onto the next emotion that needs to be learned. And there, when we can see it so clearly, it will be the only emotion in that new parallel lifetime that is what we have to work out, and it will be apparent. It will be the button that is constantly pushed. It will be the button in which we live by. But the old laws, indeed the old emotions, no longer are that which resides. It is now the next state. And when we can recognize that, indeed when we can see that, indeed when we understand that this innate emotion is not something that is coming from a new experience, all these realities are experiences that we have never experienced. It is our desire to experience them that gives us the validity of the shift in a parallel life.

But here we even find our state now fluctuating with something else that hasn't been owned. We will see it because the state will come with an indigenous emotional attitude. And when we can isolate it and know what it is, we will do the same thing. We will own it. Consciousness and energy create the nature of reality. And we will discipline our thinking by rote, and all that we think will manifest into the new paradigm and indeed into the new life shift, and the one that was obvious will not be there at all. And now we are starting to see an expansive, mobile life into which we have understood in deep contemplation that resources that are of the extraordinary can be obtained but at the cost of keeping intact our emotional body in which we use other people to get our climax. We can gain resource from that. We will understand in these shifting lifetimes, without dying, that resources that once had to go mass to mass can now be a natural environment.

And where do we go from there? So am I taking you on a fantastic journey? Let me tell you: Scientists know that in the quantum world of all potential states, all probable states, that the only thing that prevents all states simultaneously is the Observer observing and measuring the present state. And that is what you have been doing all your life. Now so what happens? What happens when there is no emotion, there is only bliss? We have a completely different quantum fixed state. How dynamic is that? The new life carries no burden, so there is no debt in the new life — there is no debt, none at all — and that immediately in this life a thought can come into this life and immediately be experienced, and then that life has undergone a quantum fluctuation. Where does then the new thought come from? If we have fulfilled the quantum state pattern of the soul and we have owned all half-spin things into a complete state, what happens then?

In Dimensional Mind[20] many years ago, I told you that there was part of your brain that under the influence of pinoline — not a drug you can take — the brain's neurons can actually unfold into the quantum field. This is what happens in that deep sleep when those itsy-bitsy, teensy-weensy watery neurons unfold into the quantum field. Well, what quantum field do you think it is unfolding into: Harry's, Jane's, or yours? (Audience: Mine.) Wonderful. Well, do you know that it is the soul who tells

20 See *The Great Architect: A Study of the Brain*, Tape 046 ed. (Yelm: Ramtha Dialogues, 1997).

the brain that in order to keep reconfiguring the body from sleep that those neurons actually unfold in the same quantum dynamic field and, because they do, the pattern of that dynamic field upon the waking state becomes the causal pattern in how we think? We cannot think, we cannot create — listen to me; this is the stinger — we cannot create new thoughts beyond the concept that these neurons have now adjusted the entire brain and thinking process. So now you ask me a question: Why do you need a Master Teacher? "Sing along with me. I'm on my way to the stars."

So now every night when the brain is repairing the body this part of the brain is dissolved right into the quantum state, and in this quantum state these neurons are programmed. So when you wake up, you can't help but think, react to the stimuli of your own life because your own life is your quantum state, an urgency for completion. Now did you get that, or did that go over here? So no one can really have any problem solved. They can be moved in the state — they can go to church, they can go to counseling — but they are never really changed. They are moved around within the quantum field.

Now what happens when, let's say, we have three neurons here. What happens, say, for example — And this is where our first fluctuation, our quantum fluctuation happened. Why did that happen? Because when this neuron — this is only a singular example — when this neuron reinforced the way the brain would perceive its own reality, consistent with the way its reality would be made, then something changed up here that forced this neuron to come back here. And the forcing of that neuron, the change was that mind had changed, the quantum state changed, so the neuron then is loaded — listen to this — reverse-firing. All of this quantum state within this neuron, this fluctuation, reverse-fires back into the brain, just like this kept reverse-firing back into the brain to keep status quo. But we have a change. This also reverse-fires.

So now why is that important? Because without reverse-firing, the brain's capacity to unfold into the quantum world means that then the very mechanism of processing thought, processing consciousness into energized thoughts of probability — the very brain itself — its very nature has to be changed. So its nature is changed by this quantum fluctuation that reverse-fires, sets an entire different tone in the brain. Now the new reality is going to be perceived exactly from the brain's point of view correctly, just like it was in the past in a former state. And if we keep changing — You see, here we have a neuron here that is now firing or dissolved down to a fluctuation in this state, a parallel shift in this state. This neuron is reverse-firing that shift into the brain's ability to have perceptional awareness of its new life.

So now not only — not only — is it vastly important that we know ourselves in the context of our mind, but in knowing ourself from that point of view we can understand the kingdom of heaven, the life that we have been given. That means if we are given life we will also die, we are a perishable dream, and that the only way we will not be perishable is to be able to understand so completely the mechanism that is the formulation of reality, and be so willing to be that mechanism to a degree that the world is detachable — you are unassailable in an emotional body — and that your greatest, greatest climax is that you have a global, humanistic, conscious, mindful God relationship of why we are here, who we are in being here, and what is expected of us.

When we know that, then the world as we see it, as we see it in others, will not be a test of our morality. We won't have to sit there and block out of our mind our temptation. That will never be necessary. We will never be able to be bought. We will never be in a position in which we allow people to convince us that they love us. We will never be in a position as a commodity. We will never, ever be in any of those positions. So the world of people, places, things, times, and events, beautiful men, beautiful women, ugly women, ugly men, color, shapes, cultures, dynamics, none are seen any longer as a preference to climax but all is owned into a great understanding. That understanding allows us to approach God.

Can we do that in one lifetime? Do we have the ability, the properties, to keep shifting life, that the more we shift, the younger we get, or we stop aging altogether, and that these shifts can last as long as two hundred years, and that we keep manifesting them? If for one New York moment you say that is a celebrity's dream, I beg to differ with you. It is a fundamental property of the immense experiences and possibilities that every human mind and indeed every human life can enjoy, and we call that, in the most meager of terms, the kingdom of heaven.

Our inability to change our life is because we think we must change it in mass to mass but, in reality, it must be changed in mind. When we do that, we indeed experience a shift. It is not until we, who raised our hands in such a truthful act, can look at our addictions and be like a great warrior and start to amass them, one by one, not in some hopeful venue but indeed in the very same way that creation itself was brought into being: God, the idea from secondary consciousness, the formulation of the image of the attitude and then becoming the image, splitting apart, and detaching yourself from the emotions. Then you are free.

The brain's capacity to reverse-fire quantum static states of infinite potential — are wholly contained — can be done here. So what is the key? Desire to pursue the adventure of knowledge. What are the ways we do that? The List now is either going to become a burdensome guilt or is going to become a tool in which one disciplines one's thoughts, and in doing that we create those shifts, those paradigm shifts, those parallel lifetimes. We are on our way. Is it possible we could never die and this very day hold, right now, the age that we have right now? Is it possible to shift quantum realities in such an unfolding way that in the unfoldment, which is called the Now — no past, no present, no future, yet past, present, and future — that if we keep unfolding in that ultimate dynamic, do we ever die? We never die. And this then is the manna of the Masters of the Far East, and this is the knowledge that they know.

So what are you going to do? Are you going to keep being spurred on by your need for climax? Are you going to halt it? Are you going to halt the activities of your life — and then you are going to take on in your mind that most seductive, analogical experience of emotion, create it analogically, put it in the fire and back off from it, and be free — or are you going to just die, because in the quantum state you already have?

And if I were you — and, of course, I am not — I would continue to study this tiny little world, and I would study it in concert to my egregious, seductive needs that I whine about, that I complain about, that I am victimized over. And, finally, for the first time, admit it: no victims, no hostages, only truth. And if you were really wise, you would do that, for contemplation is also a key in quantum observation, for in contemplation, as we contemplate that which we are, we are reverse-firing, out of our quantum state, quantum fluctuations that just by contemplation bring change, lightness, and brightness to our being. And if you don't do any of this stuff — this magnificent teaching — if you don't do any of this stuff, then you deserve, fool, to live in a frozen state in which all you are doing is consuming life until yours is gone.

To being older and wiser.
To have drank from the well of wisdom
and had thirst quenched.
To life.
So be it.
— Ramtha

Ramtha's Glossary

Analogical. Being analogical means living in the Now. It is the creative moment and is outside of time, the past, and the emotions.

Analogical mind. Analogical mind means one mind. It is the result of the alignment of primary consciousness and secondary consciousness, the Observer and the personality. The fourth, fifth, sixth, and seventh seals of the body are opened in this state of mind. The bands spin in opposite directions, like a wheel within a wheel, creating a powerful vortex that allows the thoughts held in the frontal lobe to coagulate and manifest.

Bands, the. The bands are the two sets of seven frequencies that surround the human body and hold it together. Each of the seven frequency layers of each band corresponds to the seven seals of seven levels of consciousness in the human body. The bands are the auric field that allow the processes of binary and analogical mind.

Binary mind. This term means two minds. It is the mind produced by accessing the knowledge of the human personality and the physical body without accessing our deep subconscious mind. Binary mind relies solely on the knowledge, perception, and thought processes of the neocortex and the first three seals. The fourth, fifth, sixth, and seventh seals remain closed in this state of mind.

Blue Body®. It is the body that belongs to the fourth plane of existence, the bridge consciousness, and the ultraviolet frequency band. The Blue Body® is the lord over the lightbody and the physical plane.

Blue Body® Dance. It is a discipline taught by Ramtha in which the students lift their conscious awareness to the consciousness of the fourth plane. This discipline allows the Blue Body® to be accessed and the fourth seal to be opened.

Blue Body® Healing. It is a discipline taught by Ramtha in which the students lift their conscious awareness to the consciousness of the fourth plane and the Blue Body® for the purpose of healing or changing the physical body.

Blue webs. The blue webs represent the basic structure at a subtle level of the physical body. It is the invisible skeletal structure of the physical realm vibrating at the level of ultraviolet frequency.

Body/mind consciousness. Body/mind consciousness is the consciousness that belongs to the physical plane and the human body.

Book of Life. Ramtha refers to the soul as the Book of Life, where the whole journey of involution and evolution of each individual is recorded in the form of wisdom.

C&E® = R. Consciousness and energy create the nature of reality.

C&E®. Abbreviation of Consciousness & Energy^{SM}. This is the service mark of the fundamental discipline of manifestation and the raising of consciousness taught in Ramtha's School of Enlightenment. Through this discipline the students learn to create an analogical state of mind, open up their higher seals, and create reality from the Void. A Beginning C&E® Workshop is the name of the introductory workshop for beginning students in which they learn the fundamental concepts and disciplines of Ramtha's teachings. The teachings of the Beginning C&E® Workshop can be found in *Ramtha, A Beginner's Guide to Creating Reality,* revised and expanded ed. (Yelm: JZK Publishing, a division of JZK, Inc., 2000), and in *Ramtha, Creating Personal Reality,* video ed. (Yelm: JZK Publishing, a division of JZK, Inc., 1998).

Christwalk. The Christwalk is a discipline designed by Ramtha in which the student learns to walk very slowly being acutely aware. In this discipline the students learn to manifest, with each step they take, the mind of a Christ.

Consciousness. Consciousness is the child who was born from the Void's contemplation of itself. It is the essence and fabric of all being. Everything that exists originated in consciousness and manifested outwardly through its handmaiden energy. A stream of consciousness refers to the continuum of the mind of God.

Consciousness and energy. Consciousness and energy are the dynamic force of creation and are inextricably combined. Everything that exists originated in consciousness and manifested through the modulation of its energy impact into mass.

Disciplines of the Great Work. Ramtha's School of Ancient Wisdom is dedicated to the Great Work. The disciplines of the Great Work practiced in Ramtha's School of Enlightenment are all designed in their entirety by Ramtha. These practices are powerful initiations where the student has the opportunity to apply and experience firsthand the teachings of Ramtha.

Emotional body. The emotional body is the collection of past emotions, attitudes, and electrochemical patterns that make up the brain's neuronet and define the human personality of an individual. Ramtha describes it as the seduction of the unenlightened. It is the reason for cyclical reincarnation.

Emotions. An emotion is the physical, biochemical effect of an experience. Emotions belong to the past, for they are the expression of experiences that are already known and mapped in the neuropathways of the brain.

Energy. Energy is the counterpart of consciousness. All consciousness carries with it a dynamic energy impact, radiation, or natural expression of itself. Likewise, all forms of energy carry with it a consciousness that defines it.

Enlightenment. Enlightenment is the full realization of the human person, the attainment of immortality, and unlimited mind. It is the result of raising the kundalini energy sitting at the base of the spine to the seventh seal that opens the dormant parts of the brain. When the energy penetrates the lower cerebellum and the midbrain, and the subconscious mind is opened, the individual experiences a blinding flash of light called enlightenment.

Evolution. Evolution is the journey back home from the slowest levels of frequency and mass to the highest levels of consciousness and Point Zero.

Fieldwork^{SM}. Fieldwork^{SM} is one of the fundamental disciplines of Ramtha's School of Enlightenment. The students are taught to create a symbol of something they want to know and experience and draw it on a paper card. These cards are placed with the blank side facing out on the fence rails of a large field. The students blindfold themselves and focus on their symbol, allowing their body to walk freely to find their card through the application of the law of consciousness and energy and analogical mind.

Fifth plane. The fifth plane of existence is the plane of superconsciousness and x-ray frequency. It is also known as the Golden Plane or paradise.

Fifth seal. This seal is the center of our spiritual body that connects us to the fifth plane. It is associated with the thyroid gland and with speaking and living the truth without dualism.

First plane. It refers to the material or physical plane. It is the plane of the image consciousness and Hertzian frequency. It is the slowest and densest form of coagulated consciousness and energy.

First seal. The first seal is associated with the reproductive organs, sexuality, and survival.

First three seals. The first three seals are the seals of sexuality, pain and suffering, and controlling power. These are the seals commonly at play in all of the complexities of the human drama.

Fourth plane. The fourth plane of existence is the realm of the bridge consciousness and ultraviolet frequency. This plane is described as the plane of Shiva, the destroyer of the old and creator of the new. In this plane, energy is not yet split into positive and negative polarity. Any lasting changes or healing of the physical body must be changed first at the level of the fourth plane and the Blue Body®. This plane is also called the Blue Plane, or the plane of Shiva.

Fourth seal. The fourth seal is associated with unconditional love and the thymus gland. When this seal is activated, a hormone is released that maintains the body in perfect health and stops the aging process.

God. Ramtha's teachings are an exposition of the statement, "You are God." Humanity is described as the forgotten Gods. God is different from the Void. God is the point of awareness that sprang from the Void contemplating itself. It is consciousness and energy exploring and making known the unknown potentials of the Void. It is the omnipotent and omnipresent essence of all creation.

God within. It is the Observer, the great self, the primary consciousness, the Spirit, the God within the human person.

God/man. The full realization of a human being.

God/woman. The full realization of a human being.

Gods. The Gods are technologically advanced beings from other star systems who came to Earth 455,000 years ago. These Gods manipulated the human race genetically, mixing and modifying our DNA with theirs. They are responsible for the evolution of the neocortex and used the human race as a subdued work force. Evidence of these events is recorded in the Sumerian tablets and artifacts. This term is also used to describe the true identity of humanity, the forgotten Gods.

Golden body. It is the body that belongs to the fifth plane, superconsciousness, and x-ray frequency.

Great Work. The Great Work is the practical application of the knowledge of the Schools of Ancient Wisdom. It refers to the disciplines by which the human person becomes enlightened and is transmuted into an immortal, divine being.

Hierophant. A hierophant is a master teacher who is able to manifest what they teach and initiate their students into such knowledge.

Hyperconsciousness. Hyperconsciousness is the consciousness of the sixth plane and gamma ray frequency.

Infinite Unknown. It is the frequency band of the seventh plane of existence and ultraconsciousness.

Involution. Involution is the journey from Point Zero and the seventh plane to the slowest and densest levels of frequency and mass.

JZ Knight. JZ Knight is the only person appointed by Ramtha to channel him. Ramtha refers to JZ as his beloved daughter. She was Ramaya, the eldest of the children given to Ramtha during his lifetime.

Kundalini. Kundalini energy is the life force of a person that descends from the higher seals to the base of the spine at puberty. It is a large packet of energy reserved for human evolution, commonly pictured as a coiled serpent that sits at the base of the spine. This energy is different from the energy coming out of the first three seals responsible for sexuality, pain and suffering, power and victimization. It is commonly described as the sleeping serpent or the sleeping dragon. The journey of the kundalini energy to the crown of the head is called the journey of enlightenment. This journey takes place when this serpent wakes up and starts to split and dance around the spine, ionizing the spinal fluid and changing its molecular structure. This action causes the opening of the midbrain and the door to the subconscious mind.

Life force. The life force is the Father/Mother, the Spirit, the breath of life within the person that is the platform from which the person creates its illusions, imagination, and dreams.

Life review. It is the review of the previous incarnation that occurs when the person reaches the third plane after death. The person gets the opportunity to be the Observer, the actor, and the recipient of its own actions. The unresolved issues from that lifetime that emerge at the life or light review set the agenda for the next incarnation.

Light, the. The light refers to the third plane of existence.

Lightbody. It is the same as the radiant body. It is the body that belongs to the third plane of conscious awareness and the visible light frequency band.

List, the. The List is the discipline taught by Ramtha where the student gets to write a list of items they desire to know and experience and then learn to focus on it in an analogical state of consciousness. The List is the map used to design, change, and reprogram the neuronet of the person. It is the tool that helps to bring meaningful and lasting changes in the person and their reality.

Make known the unknown. This phrase expresses the original divine mandate given to the Source consciousness to manifest and bring to conscious awareness all of the infinite potentials of the Void. This statement represents the basic intent that inspires the dynamic process of creation and evolution.

Mind. Mind is the product of streams of consciousness and energy acting on the brain creating thought-forms, holographic segments, or neurosynaptic patterns called memory. The streams of consciousness and energy are what keep the brain alive. They are its power source. A person's ability to think is what gives them a mind.

Mind of God. The mind of God comprises the mind and wisdom of every lifeform that ever lived on any dimension, in any time, or that ever will live on any planet, any star, or region of space.

Mirror consciousness. When Point Zero imitated the act of contemplation of the Void it created a mirror reflection of itself, a point of reference that made the exploration of the Void possible. It is called mirror consciousness or secondary consciousness. See **Self.**

Monkey-mind. Monkey-mind refers to the flickering, swinging mind of the personality.

Mother/Father Principle. It is the source of all life, the Father, the eternal Mother, the Void. In Ramtha's teachings, the Source and God the creator are not the same. God the creator is seen as Point Zero and primary consciousness but not as the Source, or the Void, itself.

Name-field. The name-field is the name of the large field where the discipline of Fieldwork^{SM} is practiced.

Observer. It refers to the Observer responsible for collapsing the particle/wave of quantum mechanics. It represents the great self, the Spirit, primary consciousness, the God within the human person.

Outrageous. Ramtha uses this word in a positive way to express something or someone who is extraordinary and unusual, unrestrained in action, and excessively bold or fierce.

People, places, things, times, and events. These are the main areas of human experience to which the personality is emotionally attached. These areas represent the past of the human person and constitute the content of the emotional body.

Personality, the. See **Emotional body.**

Plane of Bliss. It refers to the plane of rest where souls get to plan their next incarnations after their life reviews. It is also known as heaven and paradise where there is no suffering, no pain, no need or lack, and where every wish is immediately manifested.

Plane of demonstration. The physical plane is also called the plane of demonstration. It is the plane where the person has the opportunity to demonstrate its creative potentiality in mass and witness consciousness in material form in order to expand its emotional understanding.

Point Zero. It refers to the original point of awareness created by the Void through its act of contemplating itself. Point Zero is the original child of the Void, the birth of consciousness.

Primary consciousness. It is the Observer, the great self, the God within the human person.

Ram. Ram is a shorter version of the name Ramtha. Ramtha means the Father.

Ramaya. Ramtha refers to JZ Knight as his beloved daughter. She was Ramaya, the first one to become Ramtha's adopted child during his lifetime. Ramtha found Ramaya abandoned on the steppes of Russia. Many people gave their children to Ramtha during the march as a gesture of love and highest respect; these children were to be raised in the House of the Ram. His children grew to the great number of 133 even though he never had offspring of his own blood.

Ramtha (etymology). The name of Ramtha the Enlightened One, Lord of the Wind, means the Father. It also refers to the Ram who descended from the mountain on what is known as the Terrible Day of the Ram. "It is about that in all antiquity. And in ancient Egypt, there is an avenue dedicated to the Ram, the great conqueror. And they were wise enough to understand that whoever could walk down the avenue of the Ram could conquer the wind." The word Aram, the name of Noah's grandson, is formed from the Aramaic noun Araa — meaning earth, landmass — and the word Ramtha, meaning high. This Semitic name echoes Ramtha's descent from the high mountain, which began the great march.

Runner. A runner in Ramtha's lifetime was responsible for bringing specific messages or information. A master teacher has the ability to send runners to other people that manifest their words or intent in the form of an experience or an event.

Second plane. It is the plane of existence of social consciousness and the infrared frequency band. It is associated with pain and suffering. This plane is the negative polarity of the third plane of visible light frequency.

Second seal. This seal is the energy center of social consciousness and the infrared frequency band. It is associated with the experience of pain and suffering and is located in the lower abdominal area.

Secondary consciousness. When Point Zero imitated the act of contemplation of the Void it created a mirror reflection of itself, a point of reference that made the exploration of the Void possible. It is called mirror consciousness or secondary consciousness. See **Self.**

Self, the. The self is the true identity of the human person different from the personality. It is the transcendental aspect of the person. It refers to the secondary consciousness, the traveler in a journey of involution and evolution making known the unknown.

Sending-and-receiving. Sending-and-receiving is the name of the discipline taught by Ramtha in which the student learns to access information using the faculties of the midbrain to the exclusion of sensory perception. This discipline develops the student's psychic ability of telepathy and divination.

Seven seals. The seven seals are powerful energy centers that constitute seven levels of consciousness in the human body. The bands are the way in which the physical body is held together according to these seals. In every human being there is energy spiraling out of the first three seals or centers. The energy pulsating out of the first three seals manifests itself respectively as sexuality, pain, or power. When the upper seals are unlocked, a higher level of awareness is activated.

Seventh plane. The seventh plane is the plane of ultraconsciousness and the Infinite Unknown frequency band. This plane is where the journey of involution began. This plane was created by Point Zero when it imitated the act of contemplation of the Void and the mirror or secondary consciousness was created. A plane of existence or dimension of space and time exists between two points of consciousness. All the other planes were created by slowing down the time and frequency band of the seventh plane.

Seventh seal. This seal is associated with the crown of the head, the pituitary gland, and the attainment of enlightenment.

Shiva. The Lord God Shiva represents the Lord of the Blue Plane and the Blue Body®. Shiva is not used in reference to a singular deity from Hinduism. It is rather the representation of a state of consciousness that belongs to the fourth plane, the ultraviolet frequency band, and the opening of the fourth seal. Shiva is neither male nor female. It is an androgynous being, for the energy of the fourth plane has not yet been split into positive and negative polarity. This is an important distinction from the traditional Hindu representation of Shiva as a male deity who has a wife. The tiger skin at its feet, the trident staff, and the sun and the moon at the level of the head represent the mastery of this body over the first three seals of consciousness. The kundalini energy is pictured as fiery energy shooting from the base of the spine through the head. This is another distinction from some Hindu representations of Shiva with the serpent energy coming out at the level of the fifth seal or throat. Another symbolic image of Shiva is the long threads of dark hair and an abundance of pearl necklaces, which represent its richness of experience owned into wisdom. The quiver and bow and arrows are the agent by which Shiva shoots its powerful will and destroys imperfection and creates the new.

Sixth plane. The sixth plane is the realm of hyperconsciousness and the gamma ray frequency band. In this plane the awareness of being one with the whole of life is experienced.

Sixth seal. This seal is associated with the pineal gland and the gamma ray frequency band. The reticular formation that filters and veils the knowingness of the subconscious mind is opened when this seal is activated. The opening of the brain refers to the opening of this seal and the activation of its consciousness and energy.

Social consciousness. It is the consciousness of the second plane and the infrared frequency band. It is also called the image of the human personality and the mind of the first three seals. Social consciousness refers to the collective consciousness of human society. It is the collection of thoughts, assumptions, judgments, prejudices, laws, morality, values, attitudes, ideals, and emotions of the fraternity of the human race.

Soul. Ramtha refers to the soul as the Book of Life, where the whole journey of involution and evolution of the individual is recorded in the form of wisdom.

Subconscious mind. The seat of the subconscious mind is the lower cerebellum or reptilian brain. This part of the brain has its own independent connections to the frontal lobe and the whole of the body and has the power to access the mind of God, the wisdom of the ages.

Superconsciousness. This is the consciousness of the fifth plane and the x-ray frequency band.

Tahumo. Tahumo is the discipline taught by Ramtha in which the student learns the ability to master the effects of the natural environment — cold and heat — on the human body.

Tank field. It is the name of the large field with the labyrinth that is used for the discipline of The Tank®.

Tank®, The. It is the name given to the labyrinth used as part of the disciplines of Ramtha's School of Enlightenment. The students are taught to find the entry to this labyrinth blindfolded and move through it focusing on the Void without touching the walls or using the eyes or the senses. The objective of this discipline is to find, blindfolded, the center of the labyrinth or a room designated and representative of the Void.

Third plane. This is the plane of conscious awareness and the visible light frequency band. It is also known as the light plane and the mental plane. When the energy of the Blue Plane is lowered down to this frequency band, it splits into positive and negative polarity. It is at this point that the soul splits into two, giving origin to the phenomenon of soulmates.

Third seal. This seal is the energy center of conscious awareness and the visible light frequency band. It is associated with control, tyranny, victimization, and power. It is located in the region of the solar plexus.

Thought. Thought is different from consciousness. The brain processes a stream of consciousness, modifying it into segments — holographic pictures — of neurological, electrical, and chemical prints called thoughts. Thoughts are the building blocks of mind.

Twilight®. This term is used to describe the discipline taught by Ramtha in which the students learn to put their bodies in a catatonic state similar to deep sleep, yet retaining their conscious awareness.

Twilight® Visualization Process. It is the process used to practice the discipline of the List or other visualization formats.

Ultraconsciousness. It is the consciousness of the seventh plane and the Infinite Unknown frequency band. It is the consciousness of an ascended master.

Unknown God. The Unknown God was the single God of Ramtha's ancestors, the Lemurians. The Unknown God also represents the forgotten divinity and divine origin of the human person.

Upper four seals. The upper four seals are the fourth, fifth, sixth, and seventh seals.

Void, the. The Void is defined as one vast nothing materially, yet all things potentially. See **Mother/Father Principle.**

Yellow brain. The yellow brain is Ramtha's name for the neocortex, the house of analytical and emotional thought. The reason why it is called the yellow brain is because the neocortices were colored yellow in the original two-dimensional, caricature-style drawing Ramtha used for his teaching on the function of the brain and its processes. He explained that the different aspects of the brain in this particular drawing are exaggerated and colorfully highlighted for the sake of study and understanding. This specific drawing became the standard tool used in all the subsequent teachings on the brain.

Yeshua ben Joseph. Ramtha refers to Jesus Christ by the name Yeshua ben Joseph, following the Jewish traditions of that time.

Fig. A: The Seven Seals:
Seven Levels of Consciousness in the Human Body

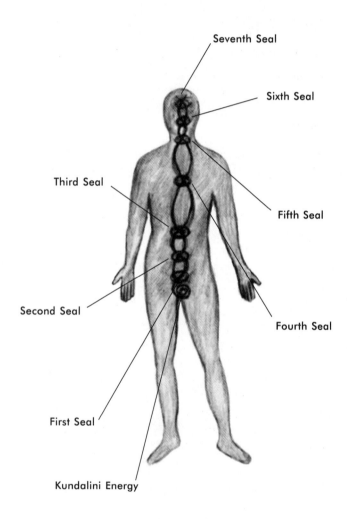

Seventh Seal

Sixth Seal

Third Seal

Fifth Seal

Second Seal

Fourth Seal

First Seal

Kundalini Energy

Fig. B: Seven Levels of Consciousness and Energy

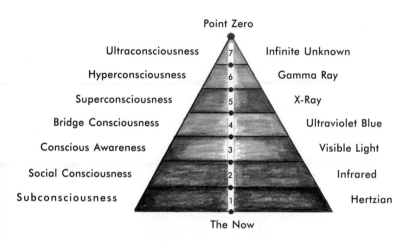

Point Zero

Ultraconsciousness	7	Infinite Unknown
Hyperconsciousness	6	Gamma Ray
Superconsciousness	5	X-Ray
Bridge Consciousness	4	Ultraviolet Blue
Conscious Awareness	3	Visible Light
Social Consciousness	2	Infrared
Subconsciousness	1	Hertzian

The Now

Fig. C: The Brain

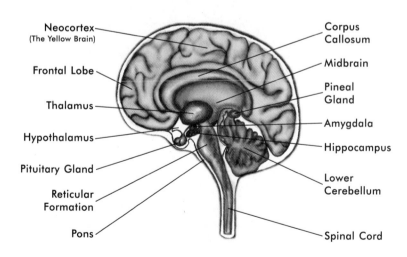

Neocortex
(The Yellow Brain)

Frontal Lobe

Thalamus

Hypothalamus

Pituitary Gland

Reticular Formation

Pons

Corpus Callosum

Midbrain

Pineal Gland

Amygdala

Hippocampus

Lower Cerebellum

Spinal Cord

Fig. D: Binary Mind — Living the Image

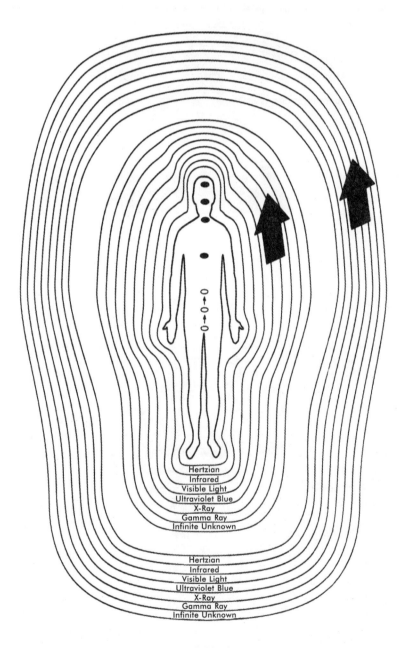

FIG. E: ANALOGICAL MIND — LIVING IN THE NOW

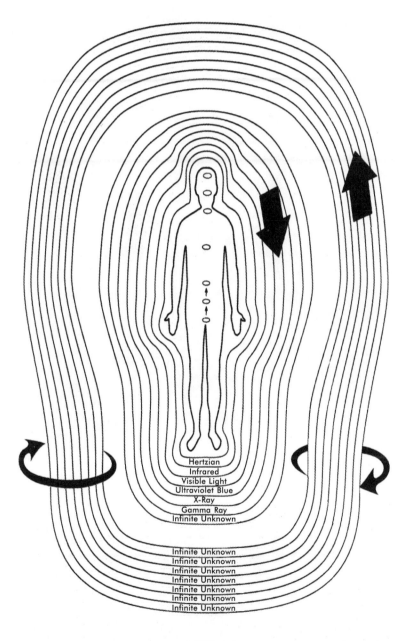

Hertzian
Infrared
Visible Light
Ultraviolet Blue
X-Ray
Gamma Ray
Infinite Unknown

Infinite Unknown
Infinite Unknown
Infinite Unknown
Infinite Unknown
Infinite Unknown
Infinite Unknown
Infinite Unknown

Fig. F: The Observer Effect and the Nerve Cell

The Observer is responsible
for collapsing the wave function of probability
into particle reality.

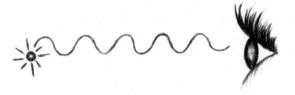

Particle   Energy wave   The Observer

The act of observation
makes the nerve cells fire and produces thought.

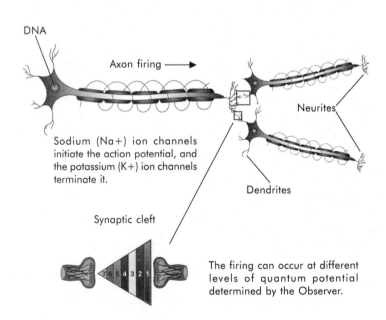

DNA

Axon firing ⟶

Neurites

Sodium (Na+) ion channels
initiate the action potential, and
the potassium (K+) ion channels
terminate it.

Dendrites

Synaptic cleft

The firing can occur at different
levels of quantum potential
determined by the Observer.

FIG. G: CELLULAR BIOLOGY AND THE THOUGHT CONNECTION

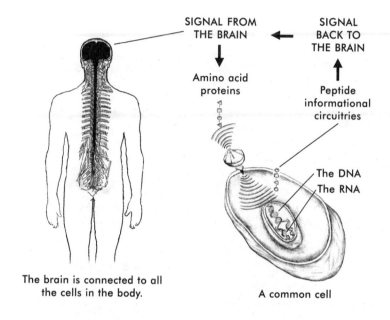

The brain is connected to all the cells in the body.

A common cell

Fig. H: Weblike Skeletal Structure of Mass

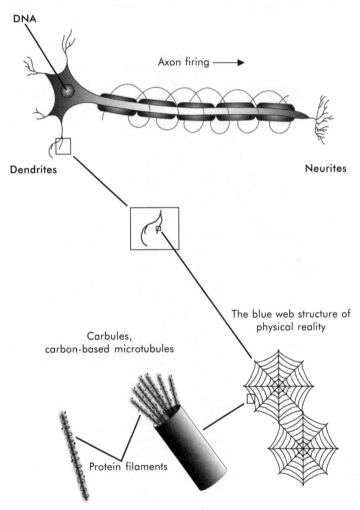

DNA

Axon firing ⟶

Dendrites

Neurites

The blue web structure of physical reality

Carbules,
carbon-based microtubules

Protein filaments

Electrons move in and through
the protein filaments.

Fig. I: The Blue Body®

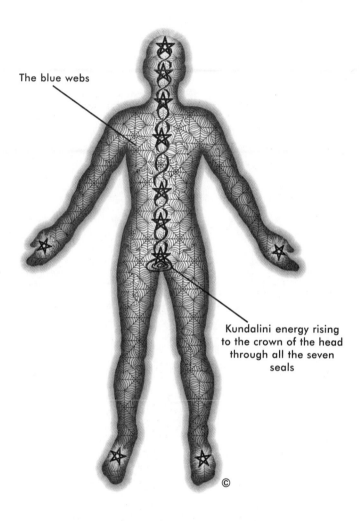

The blue webs

Kundalini energy rising
to the crown of the head
through all the seven
seals

Ramtha's School of Enlightenment
THE SCHOOL OF ANCIENT WISDOM

A Division of JZK, Inc.
P.O. Box 1210
Yelm, Washington 98597
360.458.5201
800.347.0439
www.ramtha.com
www.jzkpublishing.com